An Introduction to

Finite Projective Planes

Abraham Adrian Albert
Reuben Sandler

DOVER PUBLICATIONS, INC.
Mineola, New York

Bibliographical Note

This Dover edition, first published by Dover Publications, Inc., in 2015, is an unabridged republication of the work originally published by Holt, Rinehart and Winston, Inc., in 1968. This edition is published by special arrangement with Cengage Learning, Inc., Belmont, California.

International Standard Book Number

ISBN-13: 978-0-486-78994-1
ISBN-10: 0-486-78994-2

Manufactured in the United States by Courier Corporation
78994201 2015
www.doverpublications.com

Preface

In this book the authors have endeavored to introduce the subject of finite projective planes as it has developed during the last twenty years. It is a rich and rewarding area of mathematics, whose study uses a wide variety of techniques—geometric, algebraic, combinatorial, etc . . . . Also, it is an area containing a great many unsolved problems which require no complicated jargon to state, and which are understandable to the student who has just begun to study the subject. Finally, as is the case with much of number theory, a good deal of the study of projective planes can be taught to the student who is at an early stage in his mathematical development, and the subject can be given an elementary and self-contained presentation.

We have attempted to provide as elementary an approach as is possible consistent with the inclusion of all of the theorems we felt belonged in this book. Thus, the notion of a projective plane is introduced by way of the seven point (Fano) plane and the real projective plane, and the incidence properties of these planes are generalized to provide the axioms defining our subject. Then duality, and the axiom of Desargues are discussed along with other elementary geometric concepts. Next, the finiteness assumption is made, and several combinatorial properties of finite planes are examined in some detail, both for their own intrinsic interest and beauty, and for their utility, in leading up to the main results.

As the geometry unfolds, enough of the study of the relevant algebraic systems is covered to make the book truly self-contained. In fact, the only theorems used without being proved in the present volume are the classification theorem for finite fields, and Weddenburn's theorem on division rings. The algebraic topics introduced include loops, groups, fields, matrices, and ternary rings. Thus, the algebra introduced permits us to make a study of the collineation groups of planes, and, in particular, of the projective group of the classical finite planes defined by Galois fields.

The deviation from the classical study of the subject first occurs when the planar ternary ring is introduced as the "coordinate system" of the arbitrary projective plane. From that point on, a careful study is made of the geometric consequences of assuming more and more algebraic structure in the coordinatizing planar ternary ring of a plane until we have obtained sufficient information to prove our fundamental theorem: *A finite plane can be coordinatized by a Galois field if and only if the plane is Desarguesian.* Along the way, there is unfolded for us a set of theorems detailing the relationship between the algebraic conditions in a planar ternary ring and the geometry of the plane it coordinatizes. This lays the foundation for a more thorough study, in the mathematical literature, of groups, loops, near fields, Veblen-Weddenburn systems, etc . . . , in order to be able to understand and to possibly classify all finite projective planes in the distant future.

The book concludes with some examples of finite non-Desarguesian planes, which can easily be described in the same elementary terms as the earlier exposition.

It is the authors' contention that the subject of projective planes can be taught at virtually any undergraduate or graduate level. At the most advanced levels, of course, there is no question of this. It is a rich and interesting area of mathematics with an extensive research literature. The possibility of presenting as elementary an exposition of the subject as the present one, however, proves it to be admirably suited for teaching to undergraduates with little or no previous exposure to mathematical ideas. In fact, it seems to us to be an excellent vehicle for the introduction of a variety of mathematical concepts and theorems to the mathematical novice. Accordingly, we have attempted to write in such a way as to make the book teachable at an elementary level. We assumed virtually no knowledge or sophistication on the part of the student, and defined and discussed every algebraic system used as it became necessary. The more advanced student could speed up his study of the book by omitting those sections containing an exposition of subjects he was already familiar with. Finally, the working mathematician should be able to find here an exposition of a part of a beautiful and fascinating area of mathematics.

There are many exercises scattered throughout the book. Some require that the student complete the proof of a theorem which has been only partially proved, some ask the student to delve more deeply into the properties of the algebraic systems which are introduced, and many are simply interesting problems concerning projective planes. Taken as a whole, the problems are a significant tool for the understanding of the subject and of the mathematical tools necessary for its study.

Finally, a word about notation. In the geometric context, projective planes are always denoted by capital Greek letters, $\Pi, \Sigma, \cdots$, points by capital Roman letters, $P, Q, R, \cdots$, and lines by capital German letters, $\mathfrak{L}, \mathfrak{M}, \mathfrak{N}, \cdots$. Generally, algebraic systems are denoted by capital German letters, $\mathfrak{F}, \mathfrak{R}, \mathfrak{G}, \cdots$; and their elements by lower case Roman, $a, b, x, y, \cdots$. Functions are denoted by various symbols, $F, f, \alpha, \cdots$, depending on the context. In each chapter, the theorems, lemmas, exercises, equations, and figures are numbered from 1 on, and referred to by that number in that chapter. If, however, it is necessary, for example, to refer in Chapter IV, to Exercise 12 of Chapter III, Theorem 2 of Chapter I, or Equation 6 of Chapter II, they are referred to as Exercise 3.12, Theorem 1.2, or Equation 2.6, respectively.

A. A. Albert
R. Sandler

Chicago, Illinois
March, 1968

Contents

An Introduction to Finite Projective Planes

[1]

Elementary Results

1. Introduction

As in most areas of modern mathematics, the foundations of projective geometry can be formulated in the language of set theory. We shall begin the discussion of our subject from that point of view, since the fundamental concepts of set theory are familiar to almost everyone with recent training in elementary mathematics, and so will serve as a convenient vehicle for its introduction.

As the basic ingredients of our discussion, we shall require only the notion of a (finite or infinite) set Π of *elements* $P_1, P_2, \cdots$, together with a (finite or infinite) collection of *distinguished subsets* $\mathfrak{L}_1, \mathfrak{L}_2, \cdots$ of Π, and the concept of *inclusion*. Thus, if P is an element of Π in a distinguished subset $\mathfrak{L}$ of Π, we shall write $P \mathrm{I} \mathfrak{L}$. If P is not an element of $\mathfrak{L}$, we write $P \not\mathrm{I} \mathfrak{L}$.

We shall not be interested in every collection of subsets of every set, but shall concern ourselves only with those sets Π which, together with certain distinguished subsets $\mathfrak{L}$, satisfy the following three properties:

(a) If P and Q are *any two* distinct elements of Π, there is a *unique* distinguished subset $\mathfrak{L}$ of Π such that $P \mathrm{I} \mathfrak{L}$ and $Q \mathrm{I} \mathfrak{L}$.

(b) If $\mathfrak{L}_1$ and $\mathfrak{L}_2$ are *any two* distinguished subsets of Π, there is a *unique* element P of Π such that $P \mathrm{I} \mathfrak{L}_1$ and $P \mathrm{I} \mathfrak{L}_2$.

(c) There are four distinct elements P_1, P_2, P_3, P_4 of Π such that no three of them are elements of a common distinguished subset $\mathfrak{L}$ of Π.

Before proceeding to our general study of those sets and collections of distinguished subsets that satisfy these three conditions we shall examine certain specific examples.

2. Examples of Systems

Our first example of a mathematical system that satisfies the properties just described begins with a set labeled Π_7 and which consists of *seven* elements $P_1, P_2, P_3, P_4, P_5, P_6, P_7$. We next select *seven* distinguished subsets

$\mathfrak{L}_1, \cdots, \mathfrak{L}_7$, each consisting of *three* elements and defined as follows:

$$\begin{aligned} &\mathfrak{L}_1 = \{P_1, P_2, P_5\}, \quad &\mathfrak{L}_2 = \{P_3, P_4, P_5\}, \quad &\mathfrak{L}_3 = \{P_1, P_3, P_7\}, \\ &\mathfrak{L}_4 = \{P_2, P_4, P_7\}, \quad &\mathfrak{L}_5 = [P_1, P_4, P_6\}, \quad &\mathfrak{L}_6 = [P_2, P_3, P_6\}, \\ &\mathfrak{L}_7 = [P_5, P_6, P_7\}. \end{aligned} \tag{1}$$

Our mathematical system consists of the set Π_7 and the distinguished subsets $\mathfrak{L}_1, \cdots, \mathfrak{L}_7$ and we can verify that our inclusion properties (a), (b), and (c) all hold in a straightforward manner. For example, we can verify that no *pair* of distinct elements P_i and P_j belongs to two distinct subsets $\mathfrak{L}_q$ and $\mathfrak{L}_r$, a result requiring $(\frac{1}{2})(7)(6) = 21$ verifications, since there are 21 distinct pairs of points. Similarly the subset $\mathfrak{L}_1$ contains the points P_1, P_2, P_5 and exactly one of these is in each of the subsets $\mathfrak{L}_2, \mathfrak{L}_3, \cdots, \mathfrak{L}_7$. Thus we need 21 verifications to show that each of the 21 pairs of distinct distinguished subsets $\mathfrak{L}_q$ and $\mathfrak{L}_r$ has exactly one element common to both subsets. Finally, no three of the four distinct elements P_1, P_2, P_3, P_4 belong to any one of the subsets $\mathfrak{L}_q$; to fully check that (c) holds requires 28 such verifications.

Exercise

1. Complete a proof that the system consisting of the set Π_7 and its subsets $\mathfrak{L}_1, \cdots, \mathfrak{L}_7$ satisfies properties (a), (b), and (c).

It will be convenient to introduce a language to describe a mathematical system consisting of a set Π of elements $P_1, P_2, \cdots$ and distinguished subsets $\mathfrak{L}_1, \mathfrak{L}_2, \cdots$ satisfying properties (a), (b), and (c). We shall call each such system a *projective plane* and shall indeed say that Π *is the plane*. We shall call the elements P_i of Π the *points* of the plane and the subsets $\mathfrak{L}_q$ the *lines* of Π. We shall also refer to properties (a), (b), and (c) as the *incidence relations* for the plane Π.

Our next plane (being infinite) might possibly seem less closely related to the title of this book but it is an interesting example and worth a digression. We shall use the notation Π_R for this second plane. Its "points" will be of the following three types:

(A) The set of all ordered pairs (x, y) of real numbers x and y. We shall call these points the *ordinary* points of Π_R.

(B) The set of all real numbers (m). We shall call these points the *slope* points of Π_R.

(C) A further element Y, called the *point at infinity*.

While these names have historical significance, they have no real mathematical significance except as follows. The ordinary points of Π_R may be regarded as the points of the ordinary real Euclidean plane, that is, the plane of elementary analytic geometry. Then Π_R may be conceived of as an *extension* of the Euclidean plane $\mathfrak{E}_2$.

The distinguished subsets of Π_R will now be described. They are of three types, the first of which is the following:

(D) The subset $\mathfrak{L}_{m,k}$, consisting of all of the ordinary points (x, y) whose *coordinates* x and y satisfy the equation

$$y = (xm) + k, \tag{2}$$

together with the slope point (m). Thus, for example, $\mathfrak{L}_{1/2,3}$ consists of the slope point $(\frac{1}{2})$ and all pairs $(x, \frac{1}{2}x + 3)$. Note that $\mathfrak{L}_{m,k}$ may be thought of as consisting of the line $y = xm + k$ of $\mathfrak{E}_2$ together with the point (m) in the extended plane Π_R.

(E) The subset $\mathfrak{L}_h$, consisting of all ordinary points (x, y) with $x = h$, and the point Y. Thus $\mathfrak{L}_3$ consists of Y and all pairs $(3, y)$ for every real number y. Here we may again regard $\mathfrak{L}_h$ as consisting of the points of the line $x = h$ in $\mathfrak{E}_2$ and the point Y in its extension Π_R.

(F) The subset $\mathfrak{L}_Y$, consisting of Y and all of the slope points (m). Note that no point of $\mathfrak{L}_Y$ is in $\mathfrak{E}_2$.

The extended plane Π_R is a projective plane but the Euclidean plane itself, which does consist of points and distinguished subsets called lines, does not satisfy our properties. For, obviously, the lines of a system of parallel lines do not satisfy our *intersection* property (b). The extended Euclidean plane Π_R is really obtained as the result of joining the point Y and the slope points (m) to provide these intersections.

We are now in a position to sketch a proof of the fact that the mathematical system consisting of Π_R and its distinguished subsets $\mathfrak{L}_{m,k}$, $\mathfrak{L}_k$, and $\mathfrak{L}_Y$ does indeed satisfy properties (a), (b), and (c). Suppose that P and Q are two distinct elements of Π_R. If P and Q are both ordinary points, we write $P = (a, b)$ and $Q = (c, d)$. If $a = c$, then P and Q are both in the subset $\mathfrak{L}_a$ and no other distinguished subset contains both of them. If $a \neq c$, there is a unique line of the Euclidean plane that contains both of them and it is the set $\mathfrak{L}$ of all ordinary points (x, y) such that

$$y - b = \left(\frac{d - b}{c - a}\right)(x - a). \tag{3}$$

This line has slope m where m is given by the formula

$$m = \frac{d - b}{c - a}, \tag{4}$$

and we call the real number

$$k = b - a\left(\frac{d - b}{c - a}\right)$$

the *intercept* of this line, that is, the point $(0, k)$ on our line. Then it is easy to see that $\mathfrak{L}_{m,k}$ is the unique distinguished subset of Π_R containing both P and Q.

Our next case is that where $P = (a, b)$ is an ordinary point and $Q = (m)$ is a slope point. Then the line $\mathfrak{L}$ of the Euclidean plane whose equation is $y = (xm) + k$ contains P if $k = b - (am)$. Thus the only distinguished subset containing P and Q is the line $\mathfrak{L}_{m,k}$ containing the points $(x, (xm) + k)$ and the point (m). Finally, if $P = (a, b)$ and $Q = Y$, the set $\mathfrak{L}_a$ is the desired subset, and if P and Q are either distinct slope points, or are Y and a slope point, then our line is the line $\mathfrak{L}_Y$. Thus we have provided the essentials of a verification of the incidence property (a).

Exercise

2. Write out the details of the verification that the system Π_R described above satisfies the incidence relations (a), (b), and (c).

Observe that we are using a relation I which we are calling an incidence relation. Thus $P \mathrm{I} \mathfrak{L}$ means that the point P is in the line $\mathfrak{L}$. We shall say then that P is *incident* with $\mathfrak{L}$. However, we can also write $\mathfrak{L} \mathrm{I} P$ and read this as $\mathfrak{L}$ is incident with P. We shall actually see later that the roles of points and lines can be interchanged and a second plane called the *dual* of Π defined thereby.

3. Projective Planes

Before continuing the discussion of sets and distinguished subsets satisfying conditions (a), (b), and (c), this seems an appropriate time to introduce a less cumbersome notation for the objects we are studying. The interpretation of Π_R as the extended Euclidean plane, and the form of conditions (a) and (b) make it natural to interpret the elements of any such sets as "points" the distinguished subsets, as "lines," and the relation I as an "incidence" relation, in which case the set *together with* the distinguished subsets can reasonably be called a "plane." Accordingly, we make the following definition.

DEFINITION. A mathematical system consisting of a set Π and distinguished subsets $\mathfrak{L}_1, \mathfrak{L}_2, \cdots$ will be called a projective plane if the incidence relations (a), (b), and (c) all hold.

It should be noted here that the choice of the words "point," "line," and "plane" to describe the objects is an arbitrary one, and is not meant to suggest that the elements of an arbitrary projective plane can be interpreted as points of the ordinary *Euclidean plane.* Rather, the notation is meant to be *suggestive* of certain similarities between projective planes and the Euclidean plane, and further suggests that methods of study similar to those used to study the Euclidean plane will be fruitful in the examination of projective planes. That the choice of language is reasonably descriptive can be seen by translating conditions (a), (b), and (c) into our new notation, where $P \mathrm{I} \mathfrak{L}$ is read "P is incident with $\mathfrak{L}$," or "$\mathfrak{L}$ is incident with P."

(a′) Given any two distinct points, there is a unique line which is incident with both points,

(b′) Given any two distinct lines, there is a unique point incident with both lines,

(c′) There exist four points, no three of which are collinear, that is, incident with a single line.

Notice that the Euclidean plane satisfies (a′) and (c′), but not (b′) (since there are parallel lines in the Euclidean plane), and so the Euclidean plane is not a projective plane before its extension to Π_R. The existence of projective planes (Π_7 is an example) consisting of only a *finite* number of points (and lines), and in which each line is incident with only a *finite* number of points, is also worth noting.

4. Subplanes

A notion that is almost always definable when one is studying a mathematical system is the idea of subsystems of the systems under discussion. This is true for projective planes, and we make the following definition.

DEFINITION. Let Π be a projective plane, and let Π' be a projective plane whose points form a subset of the points of Π, and which is such that every line $\mathfrak{L}'$ of Π' is the intersection of Π' and a line $\mathfrak{L}$ of Π. Then we call Π' a subplane of Π.

As an example of the notion of subplane, we consider the plane Π_R, and let the points of Π'_R consist of the point Y, all points (x, y), with x and y rational numbers, and all points (m), where m is rational. It is easy to show that the only lines of Π_R that are incident with at least two lines of Π'_R are $\mathfrak{L}_Y$, the lines $\mathfrak{L}_{m,k}$ where m and k are both rational numbers, and the lines $\mathfrak{L}_k$ where k is rational. For example, if $\mathfrak{L}_{m,k}$ is incident with the two points (x, y), and (x_1, y_1), where x, y, x_1, y_1 are all rational numbers with

$x \neq x_1$, then as in Equation (4),

$$m = \frac{y_2 - y_1}{x_2 - x_1},$$

and

$$k = y_1 - x_1\left(\frac{y_2 - y_1}{x_2 - x_1}\right)$$

Thus, we can conclude that m and k are both rational. The remaining cases and the fact that Π'_R is a subplane of Π_R are left as an exercise.

Exercise

3. Verify that Π'_R is a subplane of Π_R.

Let us examine in more detail the idea of subplane. Another way of stating the definition is to say that Π', *a subset of the points and lines of a projective plane* Π, is a subplane of Π if the points and lines of Π' satisfy conditions (a), (b), and (c), where the incidences holding in Π' are exactly the ones holding between the points and lines when they are considered as elements of Π (that is, if P and $\mathfrak{L}$ are in Π', then $P \mathrm{I} \mathfrak{L}$ in Π' if and only if $P \mathrm{I} \mathfrak{L}$ in Π).

The reader will no doubt have noticed that this last definition has a meaning only if we agree to consider the lines of a projective plane as objects that have an existence independent of the points and are not necessarily defined as subsets of the set of points. Although we introduced the lines as subsets of the points, there is no reason why we cannot think of them as objects with their own existence that happen to be "incident with" certain of the points. To state the situation in another way, we can picture the projective plane, Π, as consisting of *two* distinct sets ρ (points) and λ (lines) and with a certain relation (I) holding between some of the elements of ρ and some of the elements of λ, and such that conditions (a′), (b′), and (c′) are satisfied.

Next, let us consider the plane Π_7, and let Π' be the subset of the *points and lines* of Π_7 consisting of the line $\mathfrak{L}_1$ and the points P_1, P_2, and P_5. Then, since $P_1 \mathrm{I} \mathfrak{L}_1$, $P_2 \mathrm{I} \mathfrak{L}_1$, $P_5 \mathrm{I} \mathfrak{L}_1$ in Π, we have the same incidences holding in Π', and indeed we can show that Π' satisfies conditions (a′) and (b′), but, of course, not (c′). To see this, notice that any two of the three points of Π' determine the unique line of Π', and since there is only one line in Π', it is true (since there can be no counterexample) that any two *distinct* lines determine a unique point in Π'. On the other hand, we could define another subset, Π'', of the points and lines of Π_7 as consisting of P_1 and the lines $\mathfrak{L}_1$, $\mathfrak{L}_3$, and $\mathfrak{L}_5$. As above, Π'' can be seen to satisfy (a′) and (b′) but not (c′).

5. Incidence Structures

Having progressed to the level of sophistication which allows us to think of lines as objects that exist in their own right rather than merely as subsets of the points, we can turn our attention to the classification of all "incidence structures" which satisfy conditions (a) and (b), but *do not satisfy* (c). Here, an "incidence structure" will mean *a set of points and a set of lines together with a relation between some of the points and some of the lines*, written $P \mathrm{I} \mathfrak{L}$. We shall prove now the following theorem, which is interesting for its own sake, and which will be quite useful later.

THEOREM 1. Let Π be an incidence structure satisfying (a′) and (b′) but not (c′). Then Π consists of one of the following types of set:

(a) the empty set;

(b_1) a single point, no lines;

(b_2) a single line, no points;

(c_1) a point P, a collection of lines $\{\mathfrak{L}_i\}$, and the incidences $P \mathrm{I} \mathfrak{L}_i$, for all i;

(c_2) a single line $\mathfrak{L}$, a collection of points $\{P_i\}$, and the incidences $P_i \mathrm{I} \mathfrak{L}$, for all i;

(d) a collection of points $\{P_i\}$ (perhaps empty), a collection of lines $\{\mathfrak{L}_j\}$ (perhaps empty), a point P, a line $\mathfrak{L}$, and the incidences $P \mathrm{I} \mathfrak{L}$, and $P \mathrm{I} \mathfrak{L}_j$ for all j, $P_i \mathrm{I} \mathfrak{L}$ for all i;

(e) a collection of points $\{P_i\}$ (perhaps empty), a collection of lines $\{\mathfrak{L}_i\}$ (perhaps empty), a point P, a line $\mathfrak{L}$, and the incidences $P_i \mathrm{I} \mathfrak{L}$ all i, $P \mathrm{I} \mathfrak{L}_i$, all i, and $P_i \mathrm{I} \mathfrak{L}_i$, all i.

Before proving the theorem, let us pause to introduce a visual aid to the study of our subject, which consists of drawing *representations* of projective planes and incidence structures in a way suggested by the geometric properties. Thus, points are *represented* by dots, and lines (that is, distinguished subsets) by joining the points of the subset. The reader must be warned not to attempt to prove anything by means of these pictures, but they do seem to aid the intuition and insight and make the subject seem more geometric than the set theoretic notation does. Thus, in Fig. 1.1, we have one *representation* of the plane Π_7, while in Fig. 1.2, we have *represented* the incidence structures of types (b_1) to (e) of Theorem 1. The representation of type (a) is left as an exercise for the reader.

Proof of Theorem 1. The proof can be broken up into several cases that cover all possibilities.

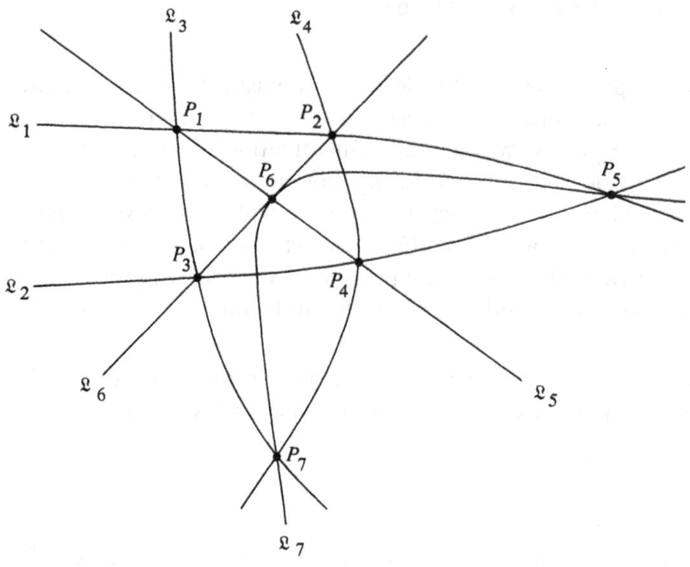
$\mathfrak{L}_3$
$\mathfrak{L}_4$
$\mathfrak{L}_1$
P_1
P_2
P_6
P_5
P_3
P_4
$\mathfrak{L}_2$
$\mathfrak{L}_6$
$\mathfrak{L}_5$
P_7
$\mathfrak{L}_7$

Fig. 1.1

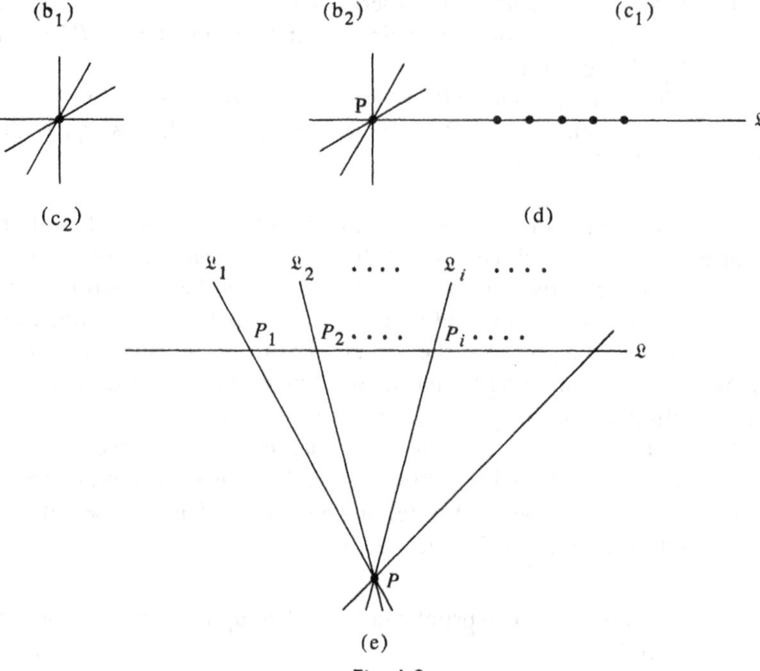
(b_1)
(b_2)
(c_1)
P
$\mathfrak{L}$
(c_2)
(d)
$\mathfrak{L}_1$ $\mathfrak{L}_2$ .... $\mathfrak{L}_i$ ....
P_1 P_2 .... P_i ....
$\mathfrak{L}$
P
(e)

Fig. 1.2

Case 1. Π has no lines. In this case, Π can have either no points or one point. If Π had two or more points, condition (a′) would imply the existence of at least one line in Π. Thus Π is either of type (a) or type (b_1).

Case 2. Π has exactly one line, $\mathfrak{L}$. If there are any points incident with $\mathfrak{L}$, there cannot exist any points in Π not incident with $\mathfrak{L}$. If there are no points incident with $\mathfrak{L}$, there could be a single point of Π not incident with $\mathfrak{L}$. Thus Π is of type (b_2), type (c_2), or perhaps of type (e), where the sets $\{P_i\}$ and $\{\mathfrak{L}_j\}$ are empty.

Case 3. Π has at least two lines in it. Then there is at most one line having more than two points incident with it. Otherwise, let $\mathfrak{L}_1$ be incident with, say, P_1, P_2, P_3, and let $\mathfrak{L}_2$ be incident with P_4, P_5, P_6. Since $\mathfrak{L}_1$ and $\mathfrak{L}_2$ have exactly one point in common by condition (b′), at least two points, say, P_1 and P_2 are on $\mathfrak{L}_1$ but not on $\mathfrak{L}_2$, while at least two points, say P_4, P_5 are on $\mathfrak{L}_2$ but not on $\mathfrak{L}_1$. But this last statement implies that no three of P_1, P_2, P_4, P_5 lie on the same line, which is condition (c′), contradicting the hypotheses of the theorem. The remainder of the analysis of case 3 is left as an exercise.

Exercise

4. Complete the proof of Theorem 1.

6. Isomorphism of Planes

One of the most important ideas in mathematics is that of isomorphism. In order to define and study the notion of isomorphism between projective planes, it is necessary to discuss two properties of functions from one set to another. If f is a function from a set S_1 to a set S_2, then f: $S_1 \to S_2$ is said to be *one-one* if

$$f(x_1) = f(x_2) \Rightarrow x_1 = x_2, \qquad (x_1, x_2 \in S_1). \tag{5}$$

Thus the function f: $R \to R$ defined by $f(x) = x + 1$ is one-one, while the function g defined by $g(x) = x^2$ is not, since $(1)^2 = (-1)^2$. The function f is said to be *onto* if, given any y in S_2, there is an x in S_1 such that $f(x) = y$.

Exercise

5. Prove that the function f: $R \to R$ defined by $f(x) = x + 1$ is one-one and onto, while the function g: $R \to R$ defined by $g(x) = x^2$ is not onto.

We can now make the following definition.

DEFINITION. The two planes Π and Σ are said to be isomorphic if there is a function $f: P \to P' = f(P)$, $\mathfrak{L} \to \mathfrak{L}' = f(\mathfrak{L})$ which maps the points of Π onto the points of Σ in a one-one fashion, which maps the lines of Π onto the lines of Σ in a one-one fashion, and such that if P is a point of Π and $\mathfrak{L}$ is a line of Π, then

$$P \mathrm{I} \mathfrak{L} \textit{ in } \Pi \qquad \textit{if and only if } f(P) \mathrm{I} f(\mathfrak{L}) \text{ in } \Sigma. \tag{6}$$

Generally speaking, isomorphic planes can be thought of as representing different ways of *naming* the elements of a single plane, and two isomorphic planes are usually referred to as being "the same" plane. As an example, consider the plane Σ whose points are $\{R_1, R_2, \cdots, R_7\}$, whose lines are $\{\mathfrak{N}_1, \cdots, \mathfrak{N}_7\}$, and whose incidences are

$$\begin{array}{lll} \mathfrak{N}_1 \mathrm{I} \{R_1, R_2, R_5\}, & \mathfrak{N}_2 \mathrm{I} \{R_3, R_4, R_5\}, & \mathfrak{N}_3 \mathrm{I} \{R_1, R_3, R_7\}, \\ \mathfrak{N}_4 \mathrm{I} \{R_2, R_4, R_7\}, & \mathfrak{N}_5 \mathrm{I} \{R_1, R_4, R_6\}, & \mathfrak{N}_6 \mathrm{I} \{R_2, R_3, R_6\}, \\ \mathfrak{N}_7 \mathrm{I} \{R_5, R_6, R_7\}. & & \end{array} \tag{7}$$

That Π and Σ are isomorphic can be seen by considering the function f defined by $f(P_i) = R_i$, $f(\mathfrak{L}_i) = \mathfrak{N}_i$, $i = 1, 2, \cdots, 7$. Then, $P_i \mathrm{I} \mathfrak{L}_j$ if and only if $R_i \mathrm{I} \mathfrak{N}_j$.

A somewhat more illuminating example is the plane Σ_1 whose points are the elements $\{S_1, \cdots, S_7\}$, whose lines are $\{\mathfrak{M}_1, \cdots, \mathfrak{M}_7\}$, and whose incidences are given by:

$$\mathfrak{M}_1 \mathrm{I} \{S_1, S_3, S_6\}, \quad \mathfrak{M}_2 \mathrm{I} \{S_1, S_4, S_5\}, \quad \mathfrak{M}_3 \mathrm{I} \{S_1, S_2, S_7\}, \quad \mathfrak{M}_4 \mathrm{I} \{S_3, S_5, S_7\},$$
$$\mathfrak{M}_5 \mathrm{I} \{S_2, S_5, S_6\}, \quad \mathfrak{M}_6 \mathrm{I} \{S_4, S_6, S_7\}, \quad \mathfrak{M}_7 \mathrm{I} \{S_2, S_3, S_4\}.$$

Exercise

6. Show, by finding an appropriate function f, that Π and Σ_1 are isomorphic.

7. Duality

Let us turn now to the study of another important idea in the subject of projective planes. To begin the discussion of the "principle of duality" for projective planes, we prove Theorem 2.

THEOREM 2. Let Π be a projective plane. Then Π contains a set of four lines, no three of which are incident with a common point.

Proof. By condition (c′), there must exist in Π a set of four points, say P_1, P_2, P_3, P_4, no three of which are incident with a common line. Condition (a′) implies that there is a unique line, say $\mathfrak{L}_1$, incident with both P_1 and P_2. Since $\mathfrak{L}_1$ is unique, there is no ambiguity in writing $\mathfrak{L}_1 = [P_1 \cdot P_2]$ to indicate that $\mathfrak{L}_1$ is the line "determined" by the distinct points P_1 and P_2. Similarly, there must exist in Π (by (a′)) the lines $\mathfrak{L}_2 = [P_3 \cdot P_4]$, $\mathfrak{L}_3 = [P_1 \cdot P_4]$, and $\mathfrak{L}_4 = [P_2 \cdot P_3]$ (see Fig. 1.3). To see that the lines $\mathfrak{L}_1$, $\mathfrak{L}_2$, $\mathfrak{L}_3$, and $\mathfrak{L}_4$ have the property stated in the theorem, consider the three lines $\mathfrak{L}_1$, $\mathfrak{L}_2$, and $\mathfrak{L}_3$. The lines $\mathfrak{L}_1$ and $\mathfrak{L}_3$ are both incident with P_1 (unique by (b′)). Thus, if $\mathfrak{L}_1$, $\mathfrak{L}_2$,

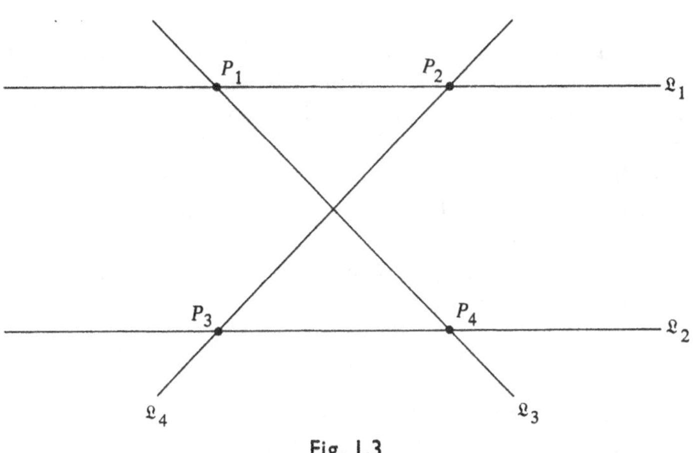

Fig. 1.3

and $\mathfrak{L}_3$ are to be incident with a common point, $\mathfrak{L}_2$ must also be incident with P_1. But if $P_1 \mathrm{I} \mathfrak{L}_2$, since we know that $P_3 \mathrm{I} \mathfrak{L}_2$ and $P_4 \mathrm{I} \mathfrak{L}_2$, we would have a contradiction to our choice of the four points P_1, P_2, P_3, and P_4. The proof that no three of the lines are incident with a common point is similar to the above proof that there is no point incident with all three of $\mathfrak{L}_1$, $\mathfrak{L}_2$, and $\mathfrak{L}_3$.

Exercise

7. Complete the proof of Theorem 2.

We can now turn to the definition of the "dual plane" of any projective plane. Before giving the definition for an arbitrary projective plane, however, let us first look at the specific plane Π_7, defined previously. Its dual plane, Π_7^*, consists of seven points, Q_1, Q_2, $\cdots$, Q_7 and seven lines,

$\mathfrak{M}_1, \cdots, \mathfrak{M}_7$ together with the incidences

$$\begin{aligned}&\mathfrak{M}_1 \mathrm{I}\{Q_1, Q_3, Q_5\}, \quad \mathfrak{M}_2 \mathrm{I}\{Q_1, Q_4, Q_6\}, \quad \mathfrak{M}_3 \mathrm{I}\{Q_2, Q_3, Q_6\},\\ &\mathfrak{M}_4 \mathrm{I}\{Q_2, Q_4, Q_5\}, \quad \mathfrak{M}_5 \mathrm{I}\{Q_1, Q_2, Q_7\}, \quad \mathfrak{M}_6 \mathrm{I}\{Q_5, Q_6, Q_7\},\\ &\mathfrak{M}_7 \mathrm{I}\{Q_3, Q_4, Q_7\}.\end{aligned} \tag{8}$$

Exercise

8. Verify that Π_7^* is a projective plane, and show that Π_7 is isomorphic with Π_7^*.

Upon comparing the incidences defining Π_7^* with the incidences defining Π_7, it becomes clear that Π_7^* was obtained by *calling* each point P_i of Π_7 a line $\mathfrak{M}_i$ of Π_7^*, considering each line $\mathfrak{L}_j$ of Π_7 a point Q_j of Π_7^*, and by using the incidences of Π_7 to define the incidences of Π_7^*, that is, $P_i \mathrm{I} \mathfrak{L}_j$ in Π_7 if and only if $Q_j \mathrm{I} \mathfrak{M}_i$ in Π_7^*. The construction of the dual plane, Π^*, of an arbitrary projective plane, Π, is defined in a completely analogous way. Let Π be a projective plane with points $\{P_i\}$ and lines $\{\mathfrak{L}_j\}$. Then the points of Π^* will consist of a set $\{Q_j\}$, while the lines of Π^* will be a set $\{\mathfrak{M}_i\}$, and we will define the incidences in Π^* by

$$Q_j \mathrm{I} \mathfrak{M}_i \text{ in } \Pi^* \qquad \text{if and only if } P_i \mathrm{I} \mathfrak{L}_j \text{ in } \Pi.$$

Another way of stating the definition of Π^* is to say that (except for changing the names to avoid confusion) the points of Π *are* the lines of Π^*, and the lines of Π *are* the points of Π^*, while the same incidences hold in Π^* as in Π. This last statement makes sense, since we have agreed to consider the points and lines of a plane as two kinds of objects whose association with each other is completely determined by the incidences holding between them. Thus we may choose to call the points "lines" and the lines "points" if we wish. That there is a reason for doing this is indicated by Theorem 3.

THEOREM 3. Let Π be a projective plane. Then the incidence structure Π^* defined above is also a projective plane.

Proof. In view of the definition of Π^*, we see that condition (a′) in Π implies that Π^* satisfies condition (b′); condition (b′) in Π implies that Π^* satisfies condition (a′); and Theorem 2 implies that Π^* satisfies condition (c′).

In general, Π and Π^* need not be isomorphic. For the "dual of the dual," $(\Pi^*)^*$, however, the situation is quite different, and it is not difficult to prove.

THEOREM 4. Let Π be any projective plane. Then Π and $(\Pi^*)^*$ are isomorphic.

Proof. Since the points and lines of $(\Pi^*)^*$ can be thought of as the lines and points, respectively, of Π^*, which, in turn, are the points and lines of Π, and since in passing from Π to Π^* and from Π^* to $(\Pi^*)^*$ we have not made any changes in the incidences, we see that Π and $(\Pi^*)^*$ are the same (that is, isomorphic).

As a corollary to Theorem 4, we have:

Every projective plane Π is (isomorphic to) a dual plane of some projective plane.

Proof. $(\Pi^*)^*$ is the dual plane of Π^*. But $(\Pi^*)^*$ is isomorphic with Π, and thus Π is isomorphic with Σ^*, where $\Sigma = \Pi^*$.

8. The Principle of Duality

With these results on dual planes, and using the theorems previously proved we can now state a "meta-theorem" in the subject, known as the "principle of duality": *Let S be any statement about projective planes, and let S* be its dual statement defined by replacing each occurrence, in S, of the word "line" by the word "point," and each "point" by "line." Then if S is a statement that is valid for* all *projective planes, then S* also holds for* all *projective planes.*

The significance of the principle of duality will be made more apparent later, when it is used to shorten substantially the proofs of several theorems. Its validity, on the other hand, can be proved by noting that, if S is true for all projective planes, then Theorem 2 and the definition of dual planes implies that S^* must be valid for every plane that is a dual plane, while the corollary to Theorem 4 then completes the proof.

9. Desargues' Configuration

Before discussing configurations in projective planes, let us define two symbols that will simplify considerably our notation. If A and B are any two distinct points of a projective plane, then we have written $[A \cdot B]$ to signify the unique line incident with both. If $\mathfrak{L}$ and $\mathfrak{M}$ are any two distinct lines of the plane, then $\mathfrak{L} \cap \mathfrak{M}$ will signify the unique point of the plane that is incident with both (sometimes called the *point of intersection* of the two lines).

Let A_1, A_2, A_3 be any three *distinct, noncollinear* points of a projective plane, Π. The points $\{A_i\}$, *together with* the lines $\mathfrak{L}_1 = [A_1 \cdot A_2]$,

$\mathfrak{L}_2 = [A_1 \cdot A_3]$, $\mathfrak{L}_3 = [A_2 \cdot A_3]$, will be called a *triangle* in Π. Since there can be no ambiguity, we shall usually refer to the "triangle" $\{A_1, A_2, A_3\}$, where it is understood that the three lines determined by the A_i also form part of the triangle. We can now define one of the more important notions in the study of projective planes.

Two triangles, $\{A_1, A_2, A_3\}$, and $\{B_1, B_2, B_3\}$ are said to be *perspective from a point P* if $P \mathrm{I} [A_i \cdot B_i]$ for $i = 1, 2, 3$. Similarly, the triangles are said to be *perspective from a line* $\mathfrak{L}$ if $([A_i \cdot A_j] \cap [B_i \cdot B_j]) \mathrm{I} \mathfrak{L}$ for $i = 1, 2$, and $j > i$, that is $([A_1 \cdot A_2] \cap [B_1 \cdot B_2]) \mathrm{I} \mathfrak{L}$, and so forth $\cdots$. These definitions are a convenient notation which allow us to introduce the following idea: A

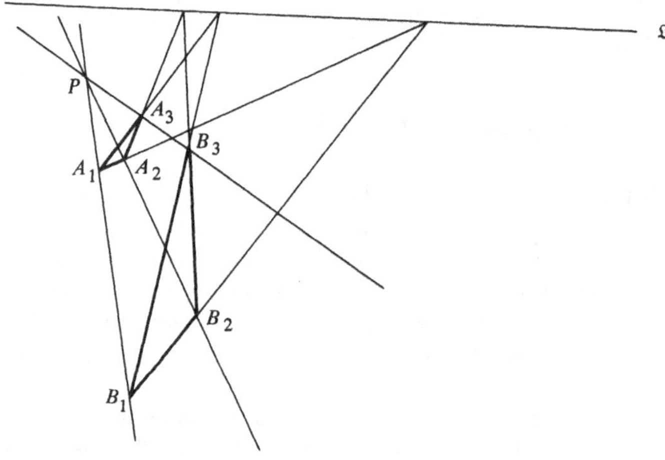

Fig. 1.4

projective plane, Π, is said to be *Desarguesian* if, whenever two triangles in Π are perspective from a point of Π, they are also perspective from some line of Π. (See Fig. 1.4.) As our study of projective planes progresses, we shall see that the definition of Desarguesian planes and planes satisfying similar geometric conditions is central to the subject. Also, we shall show that both Π_7 and Π_R are Desarguesian, while we shall construct examples of planes that are not Desarguesian. At this time we simply define the idea and begin its study in the following exercise.

Exercise

9. Let Π be a Desarguesian plane. Show that Π^* is also Desarguesian. (Hint: Show that, in Π, if two triangles are perspective from a line, they must be perspective from a point.)

[2]

Finite Planes

1. Introduction

We shall assume henceforth that all of the planes under discussion are finite. Besides discussing some of the simpler properties of finite projective planes, in this chapter we will introduce several algebraic concepts which will be most useful in our subsequent investigations.

First let us examine a second example of a finite projective plane. Let Π_{13} consist of the points $\{P_1, P_2, \cdots, P_{13}\}$, along with the lines $\{\mathfrak{L}_1, \cdots, \mathfrak{L}_{13}\}$, and the incidences

$$\begin{array}{ll} P_1 \mathrm{I} \{\mathfrak{L}_1, \mathfrak{L}_4, \mathfrak{L}_7, \mathfrak{L}_{10}\}, & P_2 \mathrm{I} \{\mathfrak{L}_2, \mathfrak{L}_5, \mathfrak{L}_8, \mathfrak{L}_{10}\} \\ P_3 \mathrm{I} \{\mathfrak{L}_3, \mathfrak{L}_6, \mathfrak{L}_9, \mathfrak{L}_{10}\}, & P_4 \mathrm{I} \{\mathfrak{L}_1, \mathfrak{L}_6, \mathfrak{L}_8, \mathfrak{L}_{11}\} \\ P_5 \mathrm{I} \{\mathfrak{L}_2, \mathfrak{L}_4, \mathfrak{L}_9, \mathfrak{L}_{11}\}, & P_6 \mathrm{I} \{\mathfrak{L}_3, \mathfrak{L}_5, \mathfrak{L}_7, \mathfrak{L}_{11}\} \\ P_7 \mathrm{I} \{\mathfrak{L}_1, \mathfrak{L}_5, \mathfrak{L}_9, \mathfrak{L}_{12}\}, & P_8 \mathrm{I} \{\mathfrak{L}_2, \mathfrak{L}_6, \mathfrak{L}_7, \mathfrak{L}_{12}\} \\ P_9 \mathrm{I} \{\mathfrak{L}_3, \mathfrak{L}_4, \mathfrak{L}_8, \mathfrak{L}_{12}\}, & P_{10} \mathrm{I} \{\mathfrak{L}_1, \mathfrak{L}_2, \mathfrak{L}_3, \mathfrak{L}_{13}\} \\ P_{11} \mathrm{I} \{\mathfrak{L}_4, \mathfrak{L}_5, \mathfrak{L}_6, \mathfrak{L}_{13}\}, & P_{12} \mathrm{I} \{\mathfrak{L}_7, \mathfrak{L}_8, \mathfrak{L}_9, \mathfrak{L}_{13}\} \\ P_{13} \mathrm{I} \{\mathfrak{L}_{10}, \mathfrak{L}_{11}, \mathfrak{L}_{12}, \mathfrak{L}_{13}\}. & \end{array} \tag{1}$$

Exercises

1. Show that Π_{13} is a projective plane.

2. Show that Π_7 and Π_{13} are not isomorphic.

2. Counting Lemmas

Now that we have two examples of finite projective planes to study, let us examine them more carefully to see if any insights can be obtained into the nature of the subject. First, notice that every line of Π_7 is incident with exactly three points of Π_7, while every point of Π_7 is incident with exactly three lines of Π_7. Similarly, every element (point or line) of Π_{13} is incident with exactly four elements (lines or points) of Π_{13}. These facts concerning Π_7 and Π_{13} lead us to the conjecture that perhaps a "similar" statement is true for any finite projective plane. That this is indeed the case is demonstrated by the results which follow.

LEMMA 1. Let Π be a finite projective plane. Then if $\mathfrak{L}_1$ and $\mathfrak{L}_2$ are any two distinct lines of Π, there exists a point P in Π not incident with $\mathfrak{L}_1$ or $\mathfrak{L}_2$.

Proof. Assume that the lemma is false. Then every point of Π is incident with either $\mathfrak{L}_1$ or $\mathfrak{L}_2$. But by condition (c′) of Section 1.3, there exist in Π four points P_1, P_2, P_3, P_4, no three of them collinear. But the only way this can happen is for two of the points, say P_1 and P_2, to be incident with $\mathfrak{L}_1$, and the other two to be incident with $\mathfrak{L}_2$ (Why?), while none of the four can be equal to $\mathfrak{L}_1 \cap \mathfrak{L}_2$ (Why?). Now consider the point $P = [P_1 \cdot P_3] \cap [P_2 \cdot P_4]$. Then P cannot be incident with $\mathfrak{L}_1$ or $\mathfrak{L}_2$ (Why?), showing that our original assumption was self-contradictory, and hence that the lemma must be true.

Exercise

3. Explain all of the points followed by (Why?) in the preceding proof.

Using Lemma 1, we can proceed to the next step.

LEMMA 2. Let Π be a finite projective plane. Then any two lines of Π are incident with the same number of points of Π.

Proof. Let $\mathfrak{L}_1$ and $\mathfrak{L}_2$ be any two distinct lines of Π, and let $R = \mathfrak{L}_1 \cap \mathfrak{L}_2$. Assume there are n additional points incident with $\mathfrak{L}_1$, $(P_1, \cdots, P_n)$, and m additional points incident with $\mathfrak{L}_2$, $(Q_1, \cdots, Q_m)$. Since any two distinct lines have exactly one common point, and $R = \mathfrak{L}_1 \cap \mathfrak{L}_2$, we know that the P's and Q's are all distinct. By Lemma 1, Π possesses a point P not incident with either $\mathfrak{L}_1$ or $\mathfrak{L}_2$. Consider now the n distinct lines $\mathfrak{M}_i = [P \cdot P_i]$ incident with P. Each of the $\mathfrak{M}_i$ must intersect $\mathfrak{L}_2$ in some point $\neq R$ (Why?), and all the points $\mathfrak{M}_i \cap \mathfrak{L}_2$ must be distinct (for if $\mathfrak{M}_i \cap \mathfrak{L}_2 = \mathfrak{M}_j \cap \mathfrak{L}_2 = Q$, then $\mathfrak{M}_i = [Q \cdot P] = \mathfrak{M}_j$, and $\mathfrak{M}_i = \mathfrak{M}_j$). But for this to be able to occur, there must be at least as many Q's as P's, so $m \geq n$. By considering the lines $\mathfrak{N}_i = [P \cdot Q_i]$, and the intersections $\mathfrak{N}_i \cap \mathfrak{L}_1$, we can reach the conclusion $n \geq m$, and this, with the inequality $m \geq n$, implies that $m = n$, which is the desired result.

As an application of Lemma 2, we can now prove Lemma 3.

LEMMA 3. Let Π be a finite projective plane. Then any two points of Π are incident with the same number of lines of Π.

Proof. Since Lemma 3 is the dual statement to Lemma 2, the principle of duality applies, and Lemma 3 must be true.

Exercise

4. Prove Lemma 3 directly, not using the principle of duality.

3. The Order of a Finite Plane

Combining the previous lemmas, and adding some new information, we come finally to the following result.

THEOREM 1. Let Π be a finite projective plane. Then there is an integer n such that

(a) Every point (line) of Π is incident with exactly $n + 1$ lines (points) of Π;

(b) Π contains exactly $n^2 + n + 1$ points (lines).

Proof. Part (a) will be proved if we can show that the number of lines with which every point is incident in the same and is the number of points with which every line is incident. To verify this, let P and $\mathfrak{L}$ be a point and line, respectively, in Π such that P is not incident with $\mathfrak{L}$. Then if there are exactly $n + 1$ lines, $\mathfrak{L}_1, \cdots, \mathfrak{L}_{n+1}$ incident with P, they must intersect with $\mathfrak{L}$ in $n + 1$ distinct (Why?) points $P_1, \cdots, P_{n+1}$, and if there were a further point, P_{n+2}, incident with $\mathfrak{L}$, the line $[P \cdot P_{n+2}]$ would be an extra line incident with P. This, together with Lemmas 2 and 3 completes the proof of part (a). To prove part (b) let P be a point of Π, and $\mathfrak{L}_1, \cdots, \mathfrak{L}_{n+1}$ the distinct lines incident with P. Then by part (a) each of the $\mathfrak{L}_i$'s is incident with n additional points (other than P). Since all the $\mathfrak{L}_i$'s are incident with P, all of these points are distinct, and furthermore, since every point is on some line incident with P (Why?), every point of Π is accounted for, and Π has $(n + 1)n = n^2 + n$ points other than P, and so, counting P, we conclude that Π has exactly $n^2 + n + 1$ points. A similar argument shows that Π has $n^2 + n + 1$ lines.

Exercises

5. Prove that if one line of a projective plane has a finite number of points on it, then the plane is finite.

6. Prove that Π has $n^2 + n + 1$ lines.

In view of Theorem 1, we are led to make the following definition:

Let Π be a finite projective plane, and let n be the integer such that Π possesses $n^2 + n + 1$ points. Then n is called the order of Π.

Observe that one of the consequences of Theorem 1 is that a *necessary* condition for a set of points and lines and a collection of incidences to form a projective plane is that there be N points and N lines, where N is of the form $n^2 + n + 1$ for some integer n. Thus, there can be a projective plane with N points, where $N = 7, 13, 21, 31, 43, 56, \cdots$. In fact, we have already constructed examples for $N = 7$ and 13 (orders 2 and 3) and there exist examples of projective planes with 21, 31, and 56 points (order 4, 5, 7) (and lines), but it is known that there is no projective plane with 43 points (order 6). One of the most important of the unsolved questions in the subject is the determination of all of those orders for which there are projective planes. A partial solution to this problem is given by the famous Bruck-Ryser theorem which is discussed in an appendix.

4. Loops and Groups

One of the beautiful features of the study of finite projective planes is the way in which it connects certain concepts of algebra with those of geometry. We shall present now an introduction of a particular concept of algebra which is of great importance for the study of our planes.

Let $\mathfrak{S}$ be a finite set of elements and let there be given a function on $\mathfrak{S}$, $\mathfrak{S}$ to $\mathfrak{S}$ which associates a *unique* element in $\mathfrak{S}$ called $x \circ y$, for every ordered pair (x, y) of elements x and y in $\mathfrak{S}$. Assume that $\mathfrak{S}$ contains an *identity* element e such that

$$x \circ e = e \circ x = x \tag{2}$$

for every x of $\mathfrak{S}$, and assume that, for every x and y of $\mathfrak{S}$ there exist unique elements a and b in $\mathfrak{S}$ such that

$$a \circ x = x \circ b = y. \tag{3}$$

Then the mathematical system consisting of the set $\mathfrak{S}$ and the "binary operation" $x \circ y$ is called a *loop.* It is customary to indicate this fact by saying that $\mathfrak{S}$ is a loop with respect to the operation $x \circ y$. When the operation is such that no confusion results, we shall simply say that $\mathfrak{S}$ is a loop. It may also be desirable to use the product notation xy for our operation $x \circ y$ and to call $\mathfrak{S}$ a multiplicative loop or to use the sum notation $x + y$ and to call $\mathfrak{S}$ an additive loop. These are used for convenience in special situations and may have special significance. However, in discussing the general loop, the notation $x \circ y$ is too bulky and we shall use the product notation.

The number of elements in the set $\mathfrak{S}$ of a loop is called the *order* of the loop. If $\mathfrak{T}$ is a subset of the set $\mathfrak{S}$ of a loop *and* the system consisting of $\mathfrak{T}$ and the binary operation of the loop $\mathfrak{S}$ applied to elements of $\mathfrak{T}$ is a loop, then the loop $\mathfrak{T}$ is called a *subloop* of $\mathfrak{S}$.

The simplest types of loops are the associative loops (called *groups*), that is, loops $\mathfrak{G}$ in which the associative law

$$(xy)z = x(yz) \tag{4}$$

holds for every x, y, z of $\mathfrak{G}$. It is easily shown that then every product $x_1 \cdots x_n$ of n factors x_i in $\mathfrak{G}$ is independent of the *grouping* of the factors. For example, if $n = 4$ this means that $x_1[(x_2x_3)x_4] = x_1[x_2(x_3x_4)] = (x_1x_2)(x_3x_4) = [(x_1x_2)x_3]x_4 = [x_1(x_2x_3)]x_4$ for any four elements x_1, x_2, x_3, x_4 of $\mathfrak{G}$.

If e is the identity element of a group $\mathfrak{G}$, the set consisting of e alone forms a subgroup of order one of $\mathfrak{G}$ with respect to the binary operation of $\mathfrak{G}$. This is the first case of what is called a cyclic group. If we make the inductive definition $g^{s+1} = g^sg$ of the (right) powers of an element g of a group $\mathfrak{G}$, then all products in $\mathfrak{G}$ with t factors all equal to g are equal to g^t. When $\mathfrak{G}$ is finite and has order n, at the most n of these powers can be distinct. Thus, if g is not the identity element e of $\mathfrak{G}$, there is an integer $m > 0$ such that $e, g, g^2, \cdots, g^{m-1}$ are distinct and $g^r = g^m$ for some r where $0 \leq r < m$. But then g^m is a product of m factors and can be written as the product $g^{m-r}g^r = g^r$. The equation $ag^r = g^r$ has the unique solution $a = e$, since $eg^r = g^r$, and so $g^{m-r} = e$ where $m - r > 0$ would contradict our definition of m unless $r = 0$. It follows that $g^m = e$ and $g^t = g^s$ for every positive integer t where $t = gm + s$, $0 \leq s < m$. It now follows that the product g^rg^s of any two powers of g with exponents r and s having values in the set $0, 1, \cdots, m - 1$ is another such power. Hence the set of powers $e, g, \cdots, g^{m-1}$ forms a subgroup of $\mathfrak{G}$ for any element g of $\mathfrak{G}$ and this subgroup has order m. We call each such subgroup of $\mathfrak{G}$ the cyclic subgroup *generated* by g and call $\mathfrak{G}$ a cyclic group if it is generated by one of its elements. If $\mathfrak{G}$ is the finite cyclic group of order n, we shall denote it $\mathfrak{C}_n$.

One method of presenting a loop is by giving a loop table, that is, a table of products. Thus, if $g_1, \cdots, g_n$ are elements of a loop $\mathfrak{S}$ of order n, the loop table for $\mathfrak{S}$ is the array

$$\begin{array}{c|cccc|} & g_1 & g_2 & \cdots & g_n \\ \hline g_1 & g_1g_1 & g_1g_2 & \cdots & g_1g_n \\ g_2 & g_2g_1 & g_2g_2 & \cdots & g_2g_n \\ & \cdot & \cdot & \cdots & \cdot \\ g_n & g_ng_1 & g_ng_2 & \cdots & g_ng_n \\ \hline \end{array},$$

where the element in the ith row and jth column of the array is the element g_k of $\mathfrak{S}$ where $g_k = g_ig_j$. The elements $g_1, \cdots, g_n$ may be permuted and the resulting array is obtained from the given array by the corresponding permutation of its rows and columns. In particular, the identity element e of $\mathfrak{G}$ may be taken as the element g_1, and so the first row of the table becomes the sequence $e, g_2, \cdots, g_n$ and the first column is the same sequence written vertically. Since the elements $g_i g_j$ and $g_i g_h$ must be distinct if $j \neq h$ by our definition of a loop, every row of a loop table contains every loop element and hence each element appears exactly once in each row. Similarly, each element appears exactly once in every column.

Finally, we define the notion of *isomorphism* between two loops (or groups). If $\mathfrak{G}$ is a loop with operation $\circ$ and $\mathfrak{T}$ is a loop with operation $*$, then $\mathfrak{G}$ and $\mathfrak{T}$ are said to be *isomorphic* if there is a *one-one* function, f, of the elements of $\mathfrak{G}$ *onto* the elements of $\mathfrak{T}$ such that

$$f(x \circ y) = f(x) * f(y) \qquad (x, y \in \mathfrak{G}). \tag{5}$$

Intuitively, this means that $\mathfrak{G}$ and $\mathfrak{T}$ are "essentially the same," except for the names of the elements and of the loop operation.

Exercises

7. Write out a loop table for the cyclic group of order six, $\mathfrak{C}_6$.

8. The definitions of a loop and a group are valid also if $\mathfrak{G}$ is an infinite set. Let $\mathfrak{G} = \mathfrak{R}$ be the set of all real numbers and define a binary operation for $\mathfrak{G}$ by $a \circ b = a + b$ for every a and b of $\mathfrak{R}$, where $+$ is the addition operation of $\mathfrak{R}$. Show that the system so defined is a group. Define a second operation by $a \circ b = ab$ where ab is the product in $\mathfrak{R}$. Show that the system consisting of $\mathfrak{R}$ and ab is not a group but that, if $\mathfrak{R}^*$ is the set obtained by deleting the zero element of $\mathfrak{R}$, then the system consisting of $\mathfrak{R}^*$ and ab is a group.

9. Let $\mathfrak{G} = \{e, g, g^2, g^3, g^4, g^5\}$, be $\mathfrak{C}_6$ (so $g^6 = e$), and let $\mathfrak{T}$ be the subset of $\mathfrak{G}$ consisting of e, g^2, and g^4. Then show that $\mathfrak{T}$ is isomorphic with $\mathfrak{C}_3$.

10. Find all the subgroups of $\mathfrak{C}_7$.

11. Find all the subgroups of $\mathfrak{C}_{12}$ and observe that they are all isomorphic with cyclic groups.

12. Find two nonisomorphic groups of order 12.

5. Collineations

The notion of a loop will be used in its full generality in later chapters, but we will now discuss certain transformations of projective planes which will be seen to make use of the idea of a group. Recall that in the last chapter we defined and discussed isomorphisms of projective planes onto other projective planes, and said that f, an isomorphism, is a 1-1 mapping of the points of Π onto the points of Σ satisfying

$$P \mathrm{I} \mathfrak{L} \Leftrightarrow f(P) \mathrm{I} f(\mathfrak{L}). \tag{6}$$

We now wish to define the "collineations" of a projective plane Π as *automorphisms* of Π, that is, isomorphisms of Π onto itself. Thus $f\colon \Pi \to \Pi$ is a *collineation* of Π if f is one-one and onto and if (6) holds for any P and $\mathfrak{L}$ in Π.

There is a "natural" binary operation "$\circ$" which can be defined on the set of collineations, $\mathfrak{G}_\Pi$, of a projective plane, Π, by $h = f \circ g$, where

$$(f \circ g)(x) \equiv f(g(x)), \tag{7}$$

where x is any point or line of Π. To see that "$\circ$" is a binary operation, note that for any $f, g \in \mathfrak{G}$, $h = f \circ g$ is a 1-1 mapping of the points of Π onto the points of Π, and the lines of Π onto the lines of Π. Also,

$$h(P) \mathrm{I} h(\mathfrak{L}) \Leftrightarrow f(g(P)) \mathrm{I} f(g(\mathfrak{L})) \Leftrightarrow g(P) \mathrm{I} g(\mathfrak{L}) \Leftrightarrow P \mathrm{I} \mathfrak{L},$$

(Why?), and so h is a collineation. In fact, the binary operation on $\mathfrak{G}$, $\mathfrak{G}$ to $\mathfrak{G}$ just defined has more significance than seems at first to be the case, as we see in the following result.

THEOREM 2. The set of collineations of a projective plane Π, together with the operation "$\circ$" form a group $\mathfrak{G}_\Pi$ called the collineation group of Π.

Proof. We shall show that all of the conditions necessary for $\mathfrak{G}_\Pi$ to form a group are satisfied by showing that $\mathfrak{G}_\Pi$ is a loop in which the associative law holds.

First, we shall show that $\mathfrak{G}_\Pi$ satisfies the associative law (4). For let f, g, and h be any three collineations in $\mathfrak{G}_\Pi$ and let P be any point of Π. Then $[f \circ (g \circ h)](P) = f[(g \circ h)(P)] = f[g(h(P))] = (f \circ g)[h(P)] = [(f \circ g) \circ h]P$. (Why are all of these equalities valid?) The same set of equalities holds for any line $\mathfrak{L}$ of Π, so we can conclude that $f \circ (g \circ h) = (f \circ g) \circ h$, the associative law.

To show that $\mathfrak{G}_\Pi$ has an identity element, consider the collineation i defined by

$$i(P) = P, \qquad i(\mathfrak{L}) = \mathfrak{L} \qquad (P, \mathfrak{L} \in \Pi). \tag{8}$$

(Why is i a collineation?) Since, for any P in Π, we have $(f \circ i)(P) = f(i(P)) = f(P) = i(f(P)) = (i \circ f)(P)$, and since a similar set of equalities holds for any $\mathfrak{L}$ in Π, we see that i is the identity collineation.

To complete the proof that $\mathfrak{G}_\Pi$ is a group, we shall first show that, for any f in $\mathfrak{G}_\Pi$, there is a unique element f^{-1}, called "the inverse of f," in $\mathfrak{G}_\Pi$ such that

$$f \circ f^{-1} = f^{-1} \circ f = i. \tag{9}$$

If f is any element of $\mathfrak{G}_\Pi$, then define a 1-1 mapping f^{-1}, by

$$\begin{aligned} f^{-1}(P) = P_1 &\Leftrightarrow f(P_1) = P, \\ f^{-1}(\mathfrak{L}) = \mathfrak{L}_1 &\Leftrightarrow f(\mathfrak{L}_1) = \mathfrak{L}. \end{aligned} \tag{10}$$

Then, for any P in Π, $f \circ f^{-1}(P) = f(f^{-1}(P)) = P$, and for any $\mathfrak{L}$ in Π, $f \circ f^{-1}(\mathfrak{L}) = \mathfrak{L}$, and so $f \circ f^{-1} = i$. Similarly, $f^{-1} \circ f = i$ (Why?), so (9) holds. Next, assume f and g are elements of $\mathfrak{G}_\Pi$, and we wish to examine the equation

$$f \circ f_1 = g. \tag{11}$$

Using the element f^{-1}, we can write $f^{-1} \circ (f \circ f_1) = f^{-1} \circ g$, and, using the associative law, we have

$$f^{-1} \circ (f \circ f_1) = (f^{-1} \circ f) \circ f_1 = i \circ f_1 = f_1 = f^{-1} \circ g.$$

Thus, we have a unique solution for f_1, that is, $f_1 = f^{-1} \circ g$ in (11). Similarly, in the equation

$$f_2 \circ f = g, \tag{12}$$

We can find a unique solution for f_2 (Why?). Thus, we have shown that $\mathfrak{G}_\Pi$ is a loop that satisfies the associative law, and the proof is complete that $\mathfrak{G}_\Pi$ is a group.

Exercises

13. Show that f^{-1} is a collineation.

14. Show that there is a unique solution for f_2 in $f_2 \circ f = g$, and give this solution explicitly in terms of f and g.

15. Let $\mathfrak{G}$ be any group. Show that for any x in $\mathfrak{G}$, there is an element x^{-1} such that $x \circ x^{-1} = x^{-1} \circ x = e$, the identity element. Does this follow from the conditions defining a loop, or is the associative law necessary?

6. The Incidence Matrix

We turn now to another characterization of finite planes that is interesting to consider. Let Π be a projective plane of order n, with points $P_1, \cdots, P_N$, and lines $\mathfrak{L}_1, \cdots, \mathfrak{L}_N$, where $N = n^2 + n + 1$. Then we shall define a square array whose size is $N \times N$, and whose entries are all zeros and ones. This array, A, is called the "incidence matrix" of Π and its entries, a_{ij}, are defined as follows:

$$a_{ij} = \begin{cases} 1 & \text{if } P_i \mathrm{I} \mathfrak{L}_j \\ 0 & \text{if } P_i \not\mathrm{I} \mathfrak{L}_j. \end{cases} \tag{13}$$

As an example, we write down A_7, the 7×7 incidence matrix for Π_7^*:

$$A_7 = \begin{pmatrix} 1 & 0 & 1 & 0 & 1 & 0 & 0 \\ 1 & 0 & 0 & 1 & 0 & 1 & 0 \\ 0 & 1 & 1 & 0 & 0 & 1 & 0 \\ 0 & 1 & 0 & 1 & 1 & 0 & 0 \\ 1 & 1 & 0 & 0 & 0 & 0 & 1 \\ 0 & 0 & 0 & 0 & 1 & 1 & 1 \\ 0 & 0 & 1 & 1 & 0 & 0 & 1 \end{pmatrix}$$

Recall that Π_7 is of order 2, and observe that each row and column of A_7 has exactly three 1s in it. Also, if any two distinct rows (or columns) are written alongside each other, they will be found to have exactly one 1 in adjacent positions. For example, the first and sixth rows:

$$\begin{matrix} 1 & 0 & 1 & 0 & 1 & 0 & 0 \\ 0 & 0 & 0 & 0 & 1 & 1 & 1. \end{matrix}$$

They each have ones in the fifth position, but in no other position do both have ones. We leave the proof that this will hold in general as an exercise.

Exercises

16. Let Π be a finite projective plane, of order n, and let A be an incidence matrix of Π.

(a) First show that each row and column of A has exactly $n + 1$ ones and n^2 zeros in it.

(b) Next, prove that if any two *distinct* rows (or columns) of A are compared, there will be *exactly one* position in which they both have ones.

17. Let $A = (a_{ij})$ be an $N \times N$ array of zeros and ones and let Π be a geometric structure with points $P_1, \cdots, P_N$, and lines $\mathfrak{L}_1, \cdots, \mathfrak{L}_N$, and incidences defined by

$$P_i \mathrm{I} \mathfrak{L}_j \qquad \text{if } a_{ij} = 1,$$

$$P_i \not\mathrm{I} \mathfrak{L}_j \qquad \text{if } a_{ij} = 0.$$

Show that, if the condition in part (b) of the previous exercise is satisfied by both the rows and columns of A, then Π satisfies conditions (a′) and (b′) of Section 1.3 defining a projective plane. What condition on A is sufficient for Π to satisfy condition (c′)?

18. Let A be the $N \times N$ incidence matrix of a plane, Π. Define the matrix B whose entries b_{ij} are given by $b_{ij} = a_{ji}$, for $i, j = 1, 2, \cdots, N$. Then show that B is the incidence matrix of the dual plane, Π^*.

For the connection between the incidence matrix of a finite projective plane and the theory of orthogonal Latin squares, see [6] in the Bibliography at the end of the book.

7. Combinatorial Results

In this section, we shall consider some results concerning collineations and subplanes. Some of the results will be needed later, and some are simply interesting and esthetically satisfying. First, let Π be a finite projective plane, and let f be a collineation of Π. Then a point P of Π is said to be a "fixed point" under f if

$$f(P) = P,$$

while a line $\mathfrak{L}$ of Π is a "fixed line" under f if

$$f(\mathfrak{L}) = \mathfrak{L}.$$

Of course, if $f = i$, the identity collineation, then every point and line of Π is fixed. As our first simple result, we have

THEOREM 3. Let Π be a projective plane, and let f be a collineation of Π. Then the fixed elements under f satisfy conditions (a′) and (b′) of Section 1.3.

Proof. Let P and Q be two distinct fixed points. Thus, $f(P) = P$, $f(Q) = Q$. Then if $\mathfrak{L} = [P \cdot Q]$, since f is a collineation, we can write:

$$f(\mathfrak{L}) = [f(P) \cdot f(Q)] = [P \cdot Q] = \mathfrak{L}.$$

Exercise

19. Why is $f(\mathfrak{L}) = [f(P) \cdot f(Q)]$? Complete the proof of Theorem 3.

The previous theorem tells us something about the collection of elements fixed by a collineation. In general, for an arbitrary collineation, there need be no fixed elements. If the collineation has certain interesting properties, much more can be said. In particular, we shall be interested in a collineation, f, satisfying the equation

$$f \circ f = i. \tag{14}$$

Such a collineation is called an *involution* if $f \neq i$.

THEOREM 4. Let Π be a finite projective plane, and let f be a collineation of Π which is an involution. Then every point of Π lies on at least one fixed line under f, and every line of Π contains at least one fixed point.

Proof. Let P be a point of Π. We distinguish two cases.

(a) P is not a fixed point of Π. Then consider the line $\mathfrak{L} = [P \cdot f(P)]$. Since $f(P) \neq P$, the line $\mathfrak{L}$ is uniquely determined. Then $f(\mathfrak{L}) = [f(P) \cdot (f \circ f)(P)]$. But $f \circ f = i$, so $(f \circ f)(P) = P$. Thus $f(\mathfrak{L}) = [f(P) \cdot P] = \mathfrak{L}$, and $\mathfrak{L}$ is a fixed line containing P. By the duality principle, every nonfixed line contains a fixed point.

(b) P is a fixed point of Π. If Q is any other fixed point of Π, then the line $\mathfrak{L} = [P \cdot Q]$ is fixed (Why?). To see that Π possesses a fixed point other than P, let $\mathfrak{M}$ be any line of Π not containing P. If $\mathfrak{M}$ is not fixed, $\mathfrak{M}$ possesses a fixed point $Q \neq P$ (Why?). If $\mathfrak{M}$ is fixed, let $\mathfrak{N}$ be yet another line of Π not containing P. If $\mathfrak{N}$ is nonfixed, then $\mathfrak{N}$ contains a fixed point. If $\mathfrak{N}$ is fixed, then $Q = \mathfrak{N} \cap \mathfrak{M}$ is fixed (Why?). Thus Π does contain a fixed point other than P in every possible case, and the theorem is proved.

Exercise

20. In the proof above, how do you know that the lines $\mathfrak{M}$ and $\mathfrak{N}$ both must exist, that is, show that, given any point P in a projective plane, there exist at least two distinct lines not containing P.

The next result to be proved deals with subplanes, but the spirit of the proof is similar to the type of argument we have been using. The theorem relates the order of a subplane to the order of the plane containing it.

THEOREM 5. Let Π be a plane of order n, and Σ a subplane (of Π) which is of order m. Then either $n = m^2$, or $n \geq m^2 + m$.

Proof. Let P be a point of Σ, and $\mathfrak{L}$ a line of Π containing P, but which is not a line of Σ. Then, since P is a point of Σ, but $\mathfrak{L}$ is not in Σ, every other point of $\mathfrak{L}$ must not be in Σ (Why?). Thus, every point of $\mathfrak{L}$ other than P can have at most one line of Σ passing through it (Why?). But $m + 1$ of the lines of Σ contain P, and thus do not intersect $\mathfrak{L}$ at any other point. Since there are m^2 more lines of Σ (Why?) and n more points of $\mathfrak{L}$ (other than P), we must have $n \geq m^2$, for we have seen that each of the remaining m^2 lines of Σ must intersect $\mathfrak{L}$ in some point $\neq P$, and the m^2 points must be distinct. Now, either $n = m^2$, or $n > m^2$, and some point of $\mathfrak{L}$ is contained in *no* lines of Σ (Why?). Let Q be such a point, that is, a point of Π incident with no lines of Σ. Now Q is incident with $n + 1$ lines of Π, and each point of Π

(hence of Σ) must lie on exactly one of these $n + 1$ lines (except Q, which lies on all of them). However, the $m^2 + m + 1$ points of Σ must lie on distinct lines of this class, for if two points of Σ were incident with one of these lines, it would be a line of Σ, which cannot occur. Thus, $n + 1 \geq m^2 + m + 1$ (Why?), which implies $n \geq m^2 + m$, as desired.

We can use the previous results to prove the following interesting theorem, which will close this chapter.

THEOREM 6. Let Π be a finite projective plane of order n, and let f be a collineation of Π, which is an involution. Then the fixed elements under f either

(a) Constitute a subplane of order $m = \sqrt{m}$, or

(b) All the fixed points (except perhaps one) lie on a single line, and all the fixed lines (except perhaps one) pass through a single point.

Proof. The proof is left as an exercise for the reader. Theorems 3 and 4 will be needed for the proof, as well as the type of reasoning used in the proof of Theorem 5.

Exercises

21. Why is $n + 1 \geq m^2 + m + 1$ in the last sentence of the proof of Theorem 5?

22. Give a proof of Theorem 6.

[3]

Field Planes

In the first two chapters, we have explored some of the simpler consequences of the conditions defining the concept of projective plane. We have also examined some specific examples of finite projective planes (Π_7 and Π_{13}), but the subject has not been systematically studied so far. For example, we have no way of knowing, at this point, whether there exist any more examples of finite projective planes at all. In this chapter, we shall undertake a systematic study of the subject of finite fields, which will remedy this lack.

Before beginning to study finite fields, however, let us step back and try to get a broad view of the subject of the chapter. The first part of the present chapter will provide an exposition of a part of the subject of finite fields. Fields are algebraic systems that can be completely characterized when finite; that is, we shall state a theorem that will classify all finite fields (see Theorem 3.1). In fact, for the geometric applications, a study of Section 3.1 and Theorem 3.1 will be sufficient, but the intervening material is included both to make this book as self-contained as possible and to increase the reader's appreciation of the interplay between algebra and geometry. The second part of the chapter will then use the finite fields (whose existence will have already been demonstrated) to construct an infinitely large class of finite projective planes. In fact, at the end of this chapter we will have examples of planes of every order $n = p^m$, where p is any prime number and m is any positive integer.

1. Fields

In the previous chapter, we studied algebraic systems that have one binary operation and satisfy certain defining conditions (groups and loops). Fields, which we shall begin to examine here, are partly characterized by having *two* binary operations defined on them. Although this would seem to make things much more complicated, finite fields are somewhat easier to discuss than groups or loops because of the connection between the two operations and because of the finiteness.

Before defining the arbitrary field, we shall look at an example. Recall that, in Exercise 2.8, we investigated certain properties of the real number system, $\mathfrak{R}$. In particular, we showed that $\mathfrak{R}$ is a group with respect to the operation $+$, and $\mathfrak{R} - \{0\}$ is a group with respect to multiplication. Also, the real numbers satisfy certain other properties that serve to connect the two operations:

$$x(y + z) = xy + xz \tag{1}$$

$$(x + y)z = xz + yz, \tag{2}$$

(distributive laws) for all real x, y, z.

Finally, the real numbers satisfy the rules

$$x + y = y + x, \tag{3}$$

$$xy = yx, \tag{4}$$

(commutative laws) for all real x and y.

Exercises

1. Show that the rational numbers, $\mathfrak{R}'$, satisfy (1) to (4), and that $\mathfrak{R}'$ forms a group with respect to addition, and $\mathfrak{R}' - \{0\}$, a group with respect to multiplication.

In the rational and real numbers, we have examples of sets on which are defined *two* binary operations, and in which certain conditions always hold. In fact, we would like to abstract these conditions and consider *finite* systems that satisfy the conditions. The reason for using these particular conditions will be clearer toward the end of this chapter, but at this point we will define the notion and examine some of the properties of fields.

Let $\mathfrak{F}$ be a finite set with two binary operations $+$, $\cdot$ defined on it. (We have chosen to denote our operations by the familiar symbols for addition and multiplication, since the properties of fields will be seen to be *analogous* to those operations as defined on the real and rational numbers.) Then $\mathfrak{F}$ will be called a *field* if $\mathfrak{F}$ satisfies the following conditions:

(a) $\mathfrak{F}$ is a group with respect to the operation $+$ (we shall denote the identity with respect to $+$, by "0"),

(b) $a + b = b + a$ for all a and b in $\mathfrak{F}$,

(c) $\mathfrak{F} - \{0\}$ is a group with respect to the operation $\cdot$ (we shall denote the identity with respect to $\cdot$ by "e," which is the element "1" in the field of rational or real numbers),

(d) $a \cdot b = b \cdot a$ for all a and b in $\mathfrak{F}$,

(e) $a \cdot (b + c) = (a \cdot b) + (a \cdot c)$ for all a, b, and c in $\mathfrak{F}$,

(f) $(a + b) \cdot c = (a \cdot c) + (b \cdot c)$ for all a, b, and c in $\mathfrak{F}$.

NOTATION. If $\mathfrak{F}$ forms a field with respect to the operations $+$, $\cdot$, then we shall denote the unique element x satisfying $a + x = 0$, by $-a$; also, we shall denote the element y satisfying $a \cdot x = e$ by a^{-1} $(a \neq 0)$.

Exercises

2. Prove that $a \cdot 0 = 0$ for all a in $\mathfrak{F}$ if $\mathfrak{F}$ is a field.

3. If a and b are elements of a field and $a \cdot b = 0$, show that either $a = 0$ or $b = 0$.

4. Show that conditions (d) and (e) imply (f).

Henceforth, we shall drop the dot in writing the multiplicative operation of a field, that is, instead of using $a \cdot b$, we shall merely write ab in order to avoid cumbersome expressions later on.

2. Prime Fields

We turn now to the investigation of a certain class of fields—fields containing a prime number p of elements (as with loops, the *order* of any finite field is defined to be the number of elements in it). This will give us an important class of fields out of which we will be able, eventually, to construct all finite fields.

First, we pause for a lemma in number theory.

LEMMA 1. Let n and m be any two positive integers. Then there is a unique pair of nonnegative integers a, b such that the following are both satisfied:

$$\begin{gathered} n = (am) + b. \\ 0 \leq b < m \end{gathered} \tag{5}$$

(Notice that b is the remainder upon dividing n by m.)

Proof. To prove the existence of such a pair a, b, we proceed by induction on n. (We may assume $m > 1$, for if $m = 1$, $n = n1 + 0$ for all n.) If $n = 1$, then $n = 0m + n$ $(n < m)$. Assume the lemma true for all integers $<n$. Then, if $n < m$, we have $n = 0m + n$. If $n \geq m$, we write $n - m = k \geq 0$. If $k = 0$, then $n = 1m + 0$. If $k > 0$, by induction we have $k = a_1m + b_1$, where $0 \leq b_1 < m$. But then $n = a_1m + m + b_1 = (a_1 + 1)m + b_1$, and $a = a_1 + 1$, $b = b_1$ is the desired pair.

Exercise

5. Show that the pair a, b is unique.

DEFINITION. In the situation of Lemma 1, we write

$$n \sim b \pmod{m} \qquad \text{where } 0 \leq b < m.$$

We can now define our class of fields—the "prime fields." For every prime number, p, let the set $\mathfrak{G}_p$ consist of the integers $\{0, 1, 2, \cdots, p - 1\}$, for example $\mathfrak{G}_2 = \{0, 1\}$, $\mathfrak{G}_3 = \{0, 1, 2\}, \cdots$. Then, in $\mathfrak{G}_p$ we shall define new operations, $\oplus$ and $\odot$, which will make $\mathfrak{G}_p$ a field.

DEFINITION. For any two elements of $\mathfrak{G}_p$, a and b define

$$a \oplus b = c$$

if

$$a + b \sim c \pmod{p},$$

and

$$a \odot b = d$$

if

$$ab \sim d \pmod{p}.$$

The set $\mathfrak{G}_p$, together with the two operations, will be designated as $GF(p)$.

To show $GF(p)$ is a field for any prime p, we must show that conditions (a) to (f) are all satisfied. To check (a), for example, we observe first that 0 satisfies the equation

$$a \oplus 0 = 0 \oplus a = a \qquad \text{for all } a \text{ in } \mathfrak{G}_p. \text{ (Why?)}$$

Thus there is an additive identity for the system. Next, we show that the associative law is satisfied, that is, that $(a_1 \oplus a_2) \oplus a_3 = a_1 \oplus (a_2 \oplus a_3)$ for all a_1, a_2, a_3 in $\mathfrak{G}_p$. But $a_1 \oplus a_2 = b_1$, where $a_1 + a_2 \sim b_1 \pmod{p}$, and $b_1 \oplus a_3 = b_2$, where $b_1 + a_3 \sim b_2 \pmod{p}$. Thus, $a_1 + a_2 = c_1p + b_1$, and $b_1 + a_3 = c_2p + b_2$, where $0 \leq b_1 < p$, and $0 \leq b_2 < p$. (Why?). So we can write $a_1 + a_2 + a_3 = (c_1 + c_2)p + b_2$ (Why?). But this shows that $a_1 + a_2 + a_3 \sim (a_1 \oplus a_2) \oplus a_3$. Similarly, we can show that $a_1 + a_2 + a_3 \sim a_1 \oplus (a_2 \oplus a_3)$ and by uniqueness, we have the desired conclusion. Next, for any a in $\mathfrak{G}_p$, we easily see that, if $b = p - a$, then $a \oplus b = b \oplus a = 0$ (Why?). Thus, we have additive inverses and, using the associative law and methods similar to those of the last part of Section 2.5, we can prove the existence of unique solutions, as required.

Exercises

6. Complete, in detail, the proof that $GF(p)$ forms a field for all prime numbers p. (Hint: "1" is the multiplicative identity.)

7. Show that, if p is not a prime, then the system $\mathfrak{G}_p$, together with the operations defined above, cannot be a field.

As a concrete example, we shall look at $GF(5)$. The set $\mathfrak{G}_5$ consists of the elements $\{0, 1, 2, 3, 4\}$. The group table for the additive operation is:

	0	1	2	3	4
0	0	1	2	3	4
1	1	2	3	4	0
2	2	3	4	0	1
3	3	4	0	1	2
4	4	0	1	2	3

The group table for the multiplicative structure is:

	1	2	3	4
1	1	2	3	4
2	2	4	1	3
3	3	1	4	2
4	4	3	2	1

Even simpler are the corresponding tables for $GF(2)$:

	0	1
0	0	1
1	1	0

	1
1	1

.

Exercise

8. Construct the tables for $GF(3)$, $GF(7)$, and $GF(11)$.

Thus we now have at our disposal a whole class of finite fields, and we shall presently show how to use finite fields to construct finite projective planes. First, however, we state without proof a classification theorem for finite fields. The proof will not be given here as it is quite long and can be found in any textbook on modern algebra, but the existence will be used in the following sections.

THEOREM 1. Any finite field is of order $n = p^m$ for some prime number p and some integer m. For any such $n = p^m$, there is a unique finite field of order n (in the sense that any two fields of order n are isomorphic).

DEFINITION. The unique finite field of order p^n will be designated $GF(p^n)$.

3. Field Planes

We shall now make use of the finite fields of Theorem 1 to define a class of finite projective planes—the *field planes*. Let $\mathfrak{F} = GF(p^n)$ be a finite field. Then, we shall define a plane Π, usually designated by $PG(2, p^n)$, as follows.

The points of Π will be represented by ordered triples of elements of $\mathfrak{F}$, *not all zero*, written horizontally:

$$P = (x_1, x_2, x_3), \tag{6}$$

where the x_i are in $\mathfrak{F}$ and are not all zero, and with the stipulation that two triples represent *the same point* if one is a nonzero constant multiple of the other, that is, $(x_1, x_2, x_3) \equiv (y_1, y_2, y_3)$ (where "$\equiv$" means the triples represent the same point) if there is an element a of $\mathfrak{F}$, $a \neq 0$, such that $x_i = ay_i$, $i = 1, 2, 3$.

The lines of Π are represented by ordered triples of elements of $\mathfrak{F}$ written vertically, and with the same stipulation on triples representing the same line:

$$\mathfrak{L} = \begin{pmatrix} x_1 \\ x_2 \\ x_3 \end{pmatrix}, \; x_i \text{ in } \mathfrak{F},\; x_i \text{ not all zero,}$$

while

$$\begin{pmatrix} x_1 \\ x_2 \\ x_3 \end{pmatrix} \equiv \begin{pmatrix} y_1 \\ y_2 \\ y_3 \end{pmatrix}$$

if $x_i = ay_i$ for some a of $\mathfrak{F}$, $a \neq 0$.

To define the incidence relation in Π, we say that the point P and the line $\mathfrak{L}$, represented respectively by (x_1, x_2, x_3) and

$$\begin{pmatrix} y_1 \\ y_2 \\ y_3 \end{pmatrix},$$

are incident if and only if

$$(x_1y_1) + (x_2y_2) + (x_3y_3) = 0. \tag{7}$$

In order to see that this definition makes sense, we must show that the definition of incidence is independent of the particular representatives (triples) of P and $\mathfrak{L}$ which are used. For, if (ax_1, ax_2, ax_3) is used for P, $(a \neq 0)$, and (y_1b, y_2b, y_3b) for $\mathfrak{L}$, $(b \neq 0)$, then $(ax_1)y_1b + (ax_2)y_2b + (ax_3)y_3b = ab(x_1y_1) + ab(x_2y_2) + ab(x_3y_3) = ab(x_1y_1 + x_2y_2 + x_3y_3)$ which is zero if and only if $x_1y_1 + x_2y_2 + x_3y_3$ is zero. Thus the definition of incidence is indeed independent of the particular representatives used.

Next, we can check conditions (a′), (b′), and (c′) of Section 1.3 to show that we have defined a projective plane. Let $P = (x_1, x_2, x_3)$, and $Q = (y_1, y_2, y_3)$ be two *distinct* points of Π. Then we would like to prove that there is a *unique* line, $\mathfrak{L}$, such that $P \mathrm{I} \mathfrak{L}$ and $Q \mathrm{I} \mathfrak{L}$. That is to say, we would like to find a triple

$$\begin{pmatrix} z_1 \\ z_2 \\ z_3 \end{pmatrix},$$

unique up to multiplication by an element of $\mathfrak{F}$, satisfying

$$\begin{aligned} &\text{(a)} \quad x_1z_1 + x_2z_2 + x_3z_3 = 0 \\ &\text{(b)} \quad y_1z_1 + y_2z_2 + y_3z_3 = 0. \end{aligned} \tag{8}$$

We proceed as follows: Not all the x_i can be zero, so we assume $x_1 \neq 0$. Then, we can solve in $\mathfrak{F}$ and get, using the properties necessarily satisfied by a finite field.

$$z_1 = -(x_2z_2 + x_3z_3)x_1^{-1}, \tag{9}$$

since x_1 has an inverse in $\mathfrak{F}$ if it is not zero. Using this equation in (8b), we get $-y_1x_2z_2x_1^{-1} - y_1x_3z_3x_1^{-1} + y_2z_2y_3z_3 = 0$, and collecting terms in z_2 and z_3, we have

$$z_2(y_2 - y_1x_2x_1^{-1}) + z_3(y_3 - y_1x_3x_1^{-1}) = 0. \tag{10}$$

Let $h_2 = y_2 - y_1x_2x_1^{-1}$, and $h_3 = y_3 - y_1x_3x_1^{-1}$. If $h_2 = h_3 = 0$, then $y_2 = x_2y_1x_1^{-1}$, $y_3 = x_3y_1x_1^{-1}$, and clearly, $y_1 = x_1y_1x_1^{-1}$, so $y_i = x_i(y_1x_1^{-1})$ for all i, a contradiction. Thus, not both h_2 and h_3 can be zero. If one is zero, say h_2, then a solution to (10) is $z_3 = 0$, z_2 arbitrary (but not zero). Then, from (9), we can solve for z_1 in terms of z_2 uniquely, $z_1 = -x_2z_2x_1^{-1}$. If we choose another value for z_2, say z_2', then $z_1' = -x_2z_2'x_1^{-1}$, and if $z_2' = az_2$, then $z_1' = az_1$ (Why?), so that the solution for $\mathfrak{L}$ is unique. The final case is $h_2 \neq 0$, $h_3 \neq 0$. Then, if z_3 is chosen arbitrarily (not zero) in $\mathfrak{F}$, from (10), we have $z_2 = z_3h_3h_2^{-1}$, and from (9) we can solve uniquely for z_1. Again, if we replace z_3 by az_3 for a not zero in $\mathfrak{F}$, z_2 is replaced by az_2, and z_1 by az_1. Thus in this final case also, the line $\mathfrak{L}$ is unique.

Properties (b′) and (c′) are dealt with in the next exercises.

Exercises

9. Write out in detail a proof that Π satisfies property (b′).

10. Show that of the four points represented by $(e, 0, 0)$, $(0, e, 0)$, $(0, 0, e)$, (e, e, e), no three are collinear.

11. Show that any point can be represented by a triple of the form (x_1, x_2, e), $(e, x_2, 0)$, or $(0, e, 0)$, (where the x_i are in $\mathfrak{F}$), and that all the points so represented are distinct. Using these representations, count the number of points in Π, and hence determine the order of Π.

12. Show that Π_7 and $PG(2, 2)$ are isomorphic.

13. Show that Π_{13} and $PG(2, 3)$ are isomorphic.

4. Matrices and Collineations of $PG(2, p^n)$

In this section we shall compute a portion of the collineation group of a field plane, but in order to do that, we need to introduce a bit of matrix theory. Here again the reader can see an example of the heavy use this geometrical subject makes of algebraic concepts.

An $n \times m$ matrix over a finite field, $\mathfrak{F}$, is an array of size $n \times m$ of elements of $\mathfrak{F}$. As examples we have the 3×3 matrix

$$A = \begin{pmatrix} a_{11} & a_{12} & a_{13} \\ a_{21} & a_{22} & a_{23} \\ a_{31} & a_{32} & a_{33} \end{pmatrix} \tag{11}$$

where the a_{ij} are in $\mathfrak{F}$, the 1×3 matrix

$$B = (b_1, b_2, b_3), \qquad (b_i \in \mathfrak{F}), \tag{12}$$

or the 3×1 matrix

$$C = \begin{pmatrix} c_1 \\ c_2 \\ c_3 \end{pmatrix}, \qquad (c_i \in \mathfrak{F}). \tag{13}$$

Thus, in our representation of the field plane over $\mathfrak{F}$, the points have been represented by classes of 1×3 matrices, and the lines of classes of 3×1 matrices.

There are certain algebraic operations which can be formally defined as follows: two matrices A_1 and A_2, *of the same dimensions*, can be composed additively by adding correspondingly placed elements of $\mathfrak{F}$, that is,

$$(a_1, a_2, a_3) \oplus (b_1, b_2, b_3) = (a_1 + b_1, a_2 + b_2, a_3 + b_3),$$

$$\begin{pmatrix} a_1 \\ a_2 \\ a_3 \end{pmatrix} \oplus \begin{pmatrix} b_1 \\ b_2 \\ b_3 \end{pmatrix} = \begin{pmatrix} a_1 + b_1 \\ a_2 + b_2 \\ a_3 + b_3 \end{pmatrix}. \tag{14}$$

Similarly, define $(a_1, a_2, a_3) \ominus (b_1, b_2, b_3) = (a_1 - b_1, a_2 - b_2, a_3 - b_3)$. Also, if A is an $n \times m$ matrix, and B is an $m \times r$ matrix, a matrix C of size $n \times r$, called the product $C = AB = (c_{ij})$, can be formed by the rule

$$c_{ij} = \sum_{k=1}^{m} a_{ik}b_{kj} = a_{i1}b_{1j} + a_{i2}b_{2j} + \cdots + a_{im}b_{mj}. \tag{15}$$

As specific examples, we compute:

$$(a_1, a_2, a_3)\begin{pmatrix} b_1 \\ b_2 \\ b_3 \end{pmatrix} = a_1b_1 + a_2b_2 + a_3b_3,$$

and

$$(a_1, a_2, a_3)\begin{pmatrix} b_{11} & b_{12} & b_{13} \\ b_{21} & b_{22} & b_{23} \\ b_{31} & b_{32} & b_{33} \end{pmatrix}$$

$$= (a_1b_{11} + a_2b_{21} + a_3b_{31}, a_1b_{12} + a_2b_{22} + a_3b_{32}, a_1b_{13} + a_2b_{23} + a_3b_{33}).$$

Exercises

14. Write out in detail the following products: AB, AC, and $(A + B)C$, where

$$A = \begin{pmatrix} a_{11} & a_{12} & a_{13} \\ a_{21} & a_{22} & a_{23} \\ a_{31} & a_{32} & a_{33} \end{pmatrix},$$

$$B = \begin{pmatrix} b_{11} & b_{12} & b_{13} \\ b_{21} & b_{22} & b_{23} \\ b_{31} & b_{32} & b_{33} \end{pmatrix},$$

$$C = \begin{pmatrix} c_1 \\ c_2 \\ c_3 \end{pmatrix}.$$

15. Prove that

$$(x_1, x_2, x_3)\begin{pmatrix} e & 0 & 0 \\ 0 & e & 0 \\ 0 & 0 & e \end{pmatrix} = (x_1, x_2, x_3),$$

and

$$\begin{pmatrix} e & 0 & 0 \\ 0 & e & 0 \\ 0 & 0 & e \end{pmatrix}\begin{pmatrix} x_1 \\ x_2 \\ x_3 \end{pmatrix} = \begin{pmatrix} x_1 \\ x_2 \\ x_3 \end{pmatrix}.$$

16. If $D = (d_1, d_2, d_3)$, and A, B, C are defined as in the previous exercise, show that $D[(AB)C] = (DA)(BC) = [(DA)B]C = D[A(BC)] = [D(AB)]C$.

17. If A and B are 1×3 matrices, and C is a 3×3 matrix, prove that

$$(AC) \oplus (BC) = (A \oplus B)C,$$

and

$$(AC) \ominus (BC) = (A \ominus B)C.$$

18. If $A = (a_1, a_2, a_3)$,

$$B = \begin{pmatrix} b_{11} & b_{12} & b_{13} \\ b_{21} & b_{22} & b_{23} \\ b_{31} & b_{32} & b_{33} \end{pmatrix},$$

and $AB = (c_1, c_2, c_3)$, show that $(da_1, da_2, da_3)B = (dc_1, dc_2, dc_3)$ for any d in $\mathfrak{F}$.

We have now developed enough of matrix theory to find some of the collineations of $\Pi = PG(2, p^n)$. Let A be a 3×3 matrix over $\mathfrak{F} = GF(p^n)$ that satisfies the condition that

$$(x_1, x_2, x_3)A = (0, 0, 0) \tag{16}$$

if and only if $x_1 = x_2 = x_3 = 0$. Then we shall construct a collineation of Π using A. If P is a point of Π, let (x_1, x_2, x_3) be a representative of P. Then $(x_1, x_2, x_3)A = (y_1, y_2, y_3)$, which is a representation of some point Q, since the y_i cannot all be zero. Thus, we have a mapping, f, of the points of Π into the points of Π. To see that f is one-one, assume $f(P) = f(P_1)$ for two distinct points P and P_1. If (x_1, x_2, x_3) and (z_1, z_2, z_3) are representations of P and P_1, respectively, then $(x_1, x_2, x_3)A = (y_1, y_2, y_3)$, and $(z_1, z_2, z_3)A = (ay_1, ay_2, ay_3)$ for some nonzero a in $\mathfrak{F}$, since $f(P) = f(P_1)$. Thus: $(a^{-1}z_1, a^{-1}z_2, a^{-1}z_3)A = (y_1, y_2, y_3)$, where $(a^{-1}z_1, a^{-1}z_2, a^{-1}z_3)$ is another representation of P_1. Therefore, by Exercise 17, $(x_1 - a^{-1}z_1, x_2 - a^{-1}z_2, x_3 - a^{-1}z_3)A = (0, 0, 0)$, or $x_i - a^{-1}z_i = 0$, $i = 1, 2, 3$ (Why?). Thus, $P = P_1$, and we can conclude that f is one-one on the points of Π.

Next, to give the whole collineation, we must define f on the lines of Π, and show that (2.6) holds for all P and $\mathfrak{L}$ in Π. Assume, for the moment, that a 3×3 matrix, A^{-1}, exists such that

$$AA^{-1} = \begin{pmatrix} e & 0 & 0 \\ 0 & e & 0 \\ 0 & 0 & e \end{pmatrix}. \tag{17}$$

The existence of such a matrix will be derived later. Then, analogous to condition (16), we have

$$A^{-1}\begin{pmatrix} y_1 \\ y_2 \\ y_3 \end{pmatrix} = \begin{pmatrix} 0 \\ 0 \\ 0 \end{pmatrix}$$

if and only if $y_1 = y_2 = y_3 = 0$. For

$$\left[AA^{-1}\begin{pmatrix} y_1 \\ y_2 \\ y_3 \end{pmatrix}\right] = (AA^{-1})\begin{pmatrix} y_1 \\ y_2 \\ y_3 \end{pmatrix} = \begin{pmatrix} y_1 \\ y_2 \\ y_3 \end{pmatrix}$$

(Why?), so if

$$A^{-1}\begin{pmatrix} y_1 \\ y_2 \\ y_3 \end{pmatrix} = \begin{pmatrix} 0 \\ 0 \\ 0 \end{pmatrix},$$

then

$$\begin{pmatrix} y_1 \\ y_2 \\ y_3 \end{pmatrix} = \begin{pmatrix} 0 \\ 0 \\ 0 \end{pmatrix}.$$

Thus, we can define $f(\mathfrak{L})$ as follows, where $\mathfrak{L}$ is a line of Π: if

$$\begin{pmatrix} y_1 \\ y_2 \\ y_3 \end{pmatrix}$$

is a representative of $\mathfrak{L}$, then

$$A^{-1}\begin{pmatrix} y_1 \\ y_2 \\ y_3 \end{pmatrix},$$

is a representative of $f(\mathfrak{L})$. As above, we can show that f is one-one on lines of Π, and now let P and $\mathfrak{L}$ be any point and line of Π represented by (x_1, x_2, x_3) and

$$\begin{pmatrix} y_1 \\ y_2 \\ y_3 \end{pmatrix},$$

respectively. Then

$$\begin{aligned} f(P)\,\mathrm{I}\,f(\mathfrak{L}) &\Leftrightarrow [(x_1, x_2, x_3)A]A^{-1}\begin{pmatrix} y_1 \\ y_2 \\ y_3 \end{pmatrix} = 0 \\ &\Leftrightarrow (x_1, x_2, x_3)[AA^{-1}]\begin{pmatrix} y_1 \\ y_2 \\ y_3 \end{pmatrix} = 0 \qquad (18) \\ &\Leftrightarrow (x_1, x_2, x_3)\begin{pmatrix} y_1 \\ y_2 \\ y_3 \end{pmatrix} = 0 \Leftrightarrow P\,\mathrm{I}\,\mathfrak{L}, \end{aligned}$$

and f is seen to be a collineation of Π.

Exercises

19. Prove that f is one-one on lines of Π.

20. Justify every implication of Equation (18).

Finally, we wish to construct some collineations of Π by showing that there exist 3×3 matrices satisfying (16) and (17). Let P_1, P_2, P_3 be *any* three noncollinear points, represented by:

$$\begin{aligned} P_1 &\equiv (x_1, x_2, x_3), \\ P_2 &\equiv (y_1, y_2, y_3), \qquad (19) \\ P_3 &\equiv (z_1, z_2, z_3), \end{aligned}$$

and consider the matrix

$$A = \begin{pmatrix} x_1 & x_2 & x_3 \\ y_1 & y_2 & y_3 \\ z_1 & z_2 & z_3 \end{pmatrix}. \qquad (20)$$

We would like to show that A has the desired properties, but we need first the following lemmas:

LEMMA 2. If P_1, P_2, P_3 are three noncollinear points with representations as given in (19), then the element defined by

$$x = x_1(y_2z_3 - y_3z_2) - x_2(y_1z_3 - y_3z_1) + x_3(y_1z_2 - y_2z_1) \tag{21}$$

cannot be zero.

Proof. All we know about P_1, P_2, and P_3 is that they are not collinear. Thus, for any line $\mathfrak{L}$ represented by

$$\begin{pmatrix} a_1 \\ a_2 \\ a_3 \end{pmatrix}, \ (a_i \text{ not all zero}),$$

the three equations

$$\begin{aligned} x_1a_1 + x_2a_2 + x_3a_3 &= 0, \\ y_1a_1 + y_2a_2 + y_3a_3 &= 0, \\ z_1a_1 + z_2a_2 + z_3a_3 &= 0, \end{aligned} \tag{22}$$

cannot all be satisfied, that is, the simultaneous linear equations (22) cannot be solved for a_i's (not all zero) in $\mathfrak{F}$. But let us try to solve anyway. First of all, since P_1 is a point, not all three of x_1, x_2, x_3 can be zero. By relabelling the points if necessary, we may assume $x_1 \neq 0$. Then

$$x_1a_1 = -(x_2a_2 + x_3a_3). \tag{23}$$

Multiplying the second and third equations by x_1, and replacing x_1a_1 by the value given in (23), we obtain

$$a_2(x_1y_2 - y_1x_2) + a_3(x_1y_3 - y_1x_3) = 0 \tag{24}$$

and

$$a_2(x_1z_2 - x_2z_1) + a_3(x_1z_3 - z_1x_3) = 0. \tag{25}$$

Now, if in (24) *both* $x_1y_2 - y_1x_2 = 0$ *and* $x_1y_3 - y_1x_3 = 0$, then any solution to (25) would automatically give a solution to (24) (Why?) so at least one is not zero, say $x_1y_2 - y_1x_2 \neq 0$. Multiplying (25) by $x_1y_2 - y_1x_2$, and using

$$a_2(x_1y_2 - y_1x_2) = -a_3(x_1y_3 - y_1x_3), \tag{26}$$

we finally obtain:

$$a_3[(x_1y_2 - y_1x_2)(x_1z_3 - z_1x_3) - (x_1y_3 - y_1x_3)(x_1z_2 - x_2z_1)] = 0. \tag{27}$$

But the left side of (27) is equal to a_3x_1x (Why?), so, since $x_1 \neq 0$, either $a_3 = 0$ or $x = 0$. If $x = 0$, then any value of a_3 is a solution of (27), and then solutions to (26) and (23) can be achieved, yielding a set of solutions to (22). But if $x \neq 0$, then $a_3 = 0$, and (26) implies $a_2 = 0$, while (23) implies $a_1 = 0$. Thus, the lemma is proved.

LEMMA 3. If A is constructed as in (20) where the points are not collinear, then $(a_1, a_2, a_3)A = (0, 0, 0)$ if and only if $a_1 = a_2 = a_3 = 0$.

Proof. The conditions for $(a_1, a_2, a_3)A = (0, 0, 0)$ can be written as follows:

$$\begin{aligned} a_1x_1 + a_2y_1 + a_3z_1 &= 0 \\ a_1x_2 + a_2y_2 + a_3z_2 &= 0 \\ a_1x_3 + a_2y_3 + a_3z_3 &= 0. \end{aligned} \tag{28}$$

Clearly, not all of x_1, y_1, z_1 are zero, or

$$P_i \mathrm{I} \begin{pmatrix} e \\ 0 \\ 0 \end{pmatrix} \qquad \text{for } i = 1, 2, 3.$$

Attempting to solve (28) for a set of a's, we get, as above, the condition that either $x = 0$, or all the a's are zero, and since $x \neq 0$ by Lemma 2, there cannot be a nonzero solution to (28).

Exercise

21. Write out in detail a proof of this lemma.

Our final step in the proof that A defines a collineation f is to show that a matrix A^{-1} exists satisfying (17). To this end, consider the matrix

$$A^{-1} = \begin{pmatrix} (y_2z_3 - y_3z_2)x^{-1}, & (x_2z_3 - x_3z_2)x^{-1}, & (x_2y_3 - x_3y_2)x^{-1} \\ (y_1z_3 - y_3z_1)x^{-1}, & (x_1z_3 - x_3z_1)x^{-1}, & (x_1y_3 - y_1x_3)x^{-1} \\ (y_1z_2 - y_2z_1)x^{-1}, & (x_1z_2 - x_2z_1)x^{-1}, & (x_1y_2 - x_2y_1)x^{-1} \end{pmatrix},$$

where x is defined by (21).

Exercise

22. Show that A and A^{-1} satisfy (17).

We summarize the results so far obtained in the following theorem.

THEOREM 2. Given any two sets of noncollinear points in a plane $\Pi = PG(2, p^n)$, P_1, P_2, P_3, and Q_1, Q_2, Q_3, there exists a collineation g of Π such that $g(P_i) = Q_i$, $i = 1, 2, 3$.

Proof. Let P_1, P_2, P_3 be any noncollinear points, then we have proved there is a collineation f constructed from the matrix

$$A = \begin{pmatrix} x_1 & x_2 & x_3 \\ y_1 & y_2 & y_3 \\ z_1 & z_2 & z_3 \end{pmatrix},$$

where the x's, y's, and z's give representations of P_1, P_2, and P_3. But observe that $(e, 0, 0)A = (x_1, x_2, x_3)$, so $f(e, 0, 0) = P_1$; similarly, $f(0, e, 0) = P_2$, and $f(0, 0, e) = P_3$. By the same methods, we can construct a collineation f_1 such that $f_1(e, 0, 0) = Q_1$, $f_1(0, e, 0) = Q_2$, $f_1(0, 0, e) = Q_3$. But since the collineations of Π form a group, f_1 has an inverse, f_1^{-1}, which satisfies $f_1^{-1}(Q_1) = (e, 0, 0)$, $f_1^{-1}(Q_2) = (0, e, 0)$, and $f_1^{-1}(Q_3) = (0, 0, e)$. We are now finished, for the collineation $g = f_1^{-1} \circ f$ has the desired property.

Now, let P_1, P_2, P_3, P_4 be any set of four points in $\Pi = PG(2, p^n)$, no three of which are collinear, and let f be the collineation (determined by the appropriate matrix) which acts as follows: $f(e, 0, 0) = P_1$, $f(0, e, 0) = P_2$, $f(0, 0, e) = P_3$. Then, since f is a collineation, there is a point P such that $f(P) = P_4$. It also follows (from the fact that f is a collineation) that no three of $(e, 0, 0)$, $(0, e, 0)$, $(0, 0, e)$, and P are collinear (Why?). Let P be represented by the matrix (x_1, x_2, x_3). Then none of the x_i can be zero. For if, say, $x_1 = 0$, then P, $(0, e, 0)$, and $(0, 0, e)$ would all be incident with the line

$$\begin{pmatrix} e \\ 0 \\ 0 \end{pmatrix}.$$

Thus we can consider the matrices

$$B = \begin{pmatrix} x_1 & 0 & 0 \\ 0 & x_2 & 0 \\ 0 & 0 & x_2 \end{pmatrix}, \qquad B^{-1} = \begin{pmatrix} x_1^{-1} & 0 & 0 \\ 0 & x_2^{-1} & 0 \\ 0 & 0 & x_3^{-1} \end{pmatrix}, \tag{29}$$

since x_i^{-1} exists for $i = 1, 2, 3$. To show, then, that B defines a collineation, we must show that $B(x, y, z) = (0, 0, 0)$ if and only if $x = y = z = 0$. But this is easily seen to be true, and by the theory previously worked out, we know that B defines a collineation, f_1, of the field plane.

The collineation f_1 defined above satisfies:

$$\begin{aligned} &f_1(e, 0, 0) = (e, 0, 0), \qquad f_1(0, e, 0) = (0, e, 0), \\ &f_1(0, 0, e) = (0, 0, e), \qquad f_1(e, e, e) = P. \end{aligned} \tag{30}$$

Using the fact that the collineations of Π form a group, we can define the collineation $g = f \circ f_1$. It is easy now to check that $g(e, 0, 0) = P_1$, $g(0, e, 0) = P_2$, $g(0, 0, e) = P_3$, and $g(e, e, e) = P_4$ (Why?). Since the P_i's were any set of four points, no three collinear, we have demonstrated the truth of the following theorem.

THEOREM 3. Given a set of four points $\{P_i\}$ in $\Pi = PG(2, p^n)$ no three of which are collinear, and another set of four points $\{Q_i\}$ in Π, no three collinear, then there exists a collineation, g, of Π such that $g(P_i) = Q_i$, for $i = 1, 2, 3, 4$.

Proof. The theorem is an extension of, and its proof analogous to, that of Theorem 2.

Exercise

23. Write out a proof of the theorem.

5. Analytic Geometry—Coordinates

We have seen how it is possible to use a field to define a projective plane, and we have been able to study that plane to some extent in the previous section. Now, let us step back a little and try to get another view of what has been done, by examining somewhat more closely the points and lines of $\Pi = PG(2, p^n)$. As we have remarked previously, there are three kinds of points: type (1) consists of points that can be represented by elements (x, y, e), where x and y are in $GF(p^n) = \mathfrak{F}$; type (2) points can be represented by elements $(e, x, 0)$, x in $\mathfrak{F}$; and there is one point of type (3), represented by $(0, e, 0)$. There are $(p^n)^2$ points of type (1) and p^n of type (2) (Why?). The lines can be similarly divided into points represented by

$$\begin{pmatrix} m \\ -e \\ k \end{pmatrix}, \quad \begin{pmatrix} -e \\ 0 \\ k \end{pmatrix}, \quad \begin{pmatrix} 0 \\ 0 \\ e \end{pmatrix},$$

for m and k in $\mathfrak{F}$.

Consider now a point P of type (1) and a line $\mathfrak{L}$ of type (1), (x, y, e), and

$$\begin{pmatrix} m \\ -e \\ k \end{pmatrix},$$

respectively. Then

$$P \mathrm{I} \mathfrak{L} \Leftrightarrow xm - y + k = 0 \tag{31}$$

(Why?). That is to say,

$$P \mathrm{I} \mathfrak{L} \Leftrightarrow y = xm + k. \tag{32}$$

Thus, we can see that the line $\mathfrak{L}$ represented by

$$\begin{pmatrix} m \\ -e \\ k \end{pmatrix}$$

consists of all points of the form $(x, xm + k, e)$, along with the point $(e, m, 0)$. Similarly, the line of type

$$\begin{pmatrix} -e \\ 0 \\ k \end{pmatrix}$$

consists of all points represented by (k, y, e), and the point $(0, e, 0)$. Finally the line represented by

$$\begin{pmatrix} 0 \\ 0 \\ e \end{pmatrix}$$

contains all points of types (2) and (3).

But what is the import of this last paragraph? The points of type (1) can be represented by *ordered pairs* (x, y) of elements of $\mathfrak{F}$, the points of type (2), by *elements* of $\mathfrak{F}$, and the point of type (3) is a special element. Also, the lines can be interpreted in this way. But we are left then with a representation of $PG(2, p^n)$ which is completely analogous to the construction of $\Pi_{\mathfrak{R}}$ (see Section 1.2), except that we are using a finite field $GF(p^n)$ instead of the field of real numbers as the coordinate field. Thus, by omitting the line

$$\begin{pmatrix} 0 \\ 0 \\ e \end{pmatrix}$$

and its points, we are left with *an analogue of the Euclidean plane*, in which points are represented by ordered pairs (x, y), and lines by equations of the form $y = xm + k$, and $x = k$, for the various values of m and k.

In Chapter 4, we shall follow this up and see how to introduce "coordinates" in an arbitrary finite projective plane, where, except for a special point, the points will be represented by ordered pairs or single elements of our coordinatizing set and the same holds for lines. Certain properties of the coordinatizing set will then be seen to be consequences of the geometrical properties of the plane.

[4]

Coordinates in an Arbitrary Plane

In Chapter 3, we saw how to use a field to define a projective plane, and how the field could be interpreted as a "coordinate system" in the plane. In this chapter, we shall be concerned with the problem of introducing an algebraic coordinate system in any plane, and with investigating some of the properties of these coordinate systems.

1. Naming the Points and Lines

Let Π be a finite projective plane of order n, and let $\{X, Y, Q, I\}$ be an ordered set of four points, no three of which are collinear. Such a set will be called a *coordinatizing quadrangle* of Π. We choose a set $\mathfrak{R}$ containing any n distinct symbols, with the stipulation that $\mathfrak{R}$ contain among its elements the symbols "0," and "1." This must always be possible since $n \geq 2$. Finally, let $\mathfrak{L}_1 = [X \cdot Q]$, $\mathfrak{L}_2 = [Y \cdot Q]$, $\mathfrak{L} = [X \cdot Y]$, and $\mathfrak{L}' = [Q \cdot I]$. Now we can begin naming the elements of Π by using the elements of $\mathfrak{R}$.

To each point P of $\mathfrak{L}'$, except $\mathfrak{L}' \cap \mathfrak{L}$, assign a distinct element of $\mathfrak{R}$, $\alpha(P)$, with the provision that $\alpha(Q) = 0$, and $\alpha(I) = 1$. Thus, the mapping $P \rightarrow \alpha(P)$ is a one-one correspondence between the points of $\mathfrak{L}'$ (other than $\mathfrak{L}' \cap \mathfrak{L}$) and the set $\mathfrak{R}$. Now, let P be any point of Π, but not on $\mathfrak{L}$. Then, $[P \cdot Y] \cap \mathfrak{L}' \neq \mathfrak{L}' \cap \mathfrak{L}$ (Why?), so $\alpha([P \cdot Y] \cap \mathfrak{L}') = a$. Similarly, $\alpha([P \cdot X] \cap \mathfrak{L}') = b$. Then we shall give the point P *the name* (a, b) (see Fig. 4.1). Observe that each point not on $\mathfrak{L}$ has been named uniquely in this way (Why?) and that each ordered pair of elements of $\mathfrak{R}$ has been made to correspond to such a point (Why?). Also, notice that, if $P \mathrel{I} \mathfrak{L}'$ and $\alpha(P) = a$, then P has been named (a, a). Thus, Q has been named $(0, 0)$, and I, $(1, 1)$.

Exercise

1. Show that, if $\mathfrak{M} = [Y \cdot (x, x)]$ is any line through Y distinct from $\mathfrak{L}$, it consists of Y and all points of the form (x, a), for a in $\mathfrak{R}$. What similar statement can you make about the lines containing the point X?

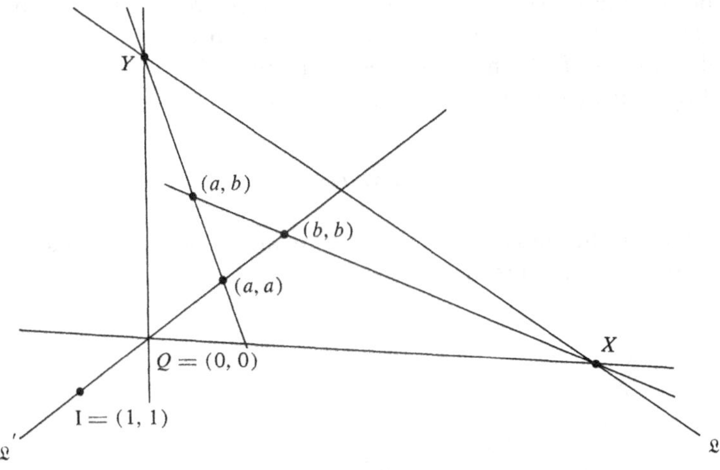

Fig. 4.1

To finish naming the points of Π, we consider the line $[Y \cdot I]$. Except for Y itself, it consists of all the points of the form $(1, x)$, with x in $\mathfrak{R}$. The point $([Q \cdot (1, m)] \cap \mathfrak{L})$ is named (m), and Y is simply called Y. Thus, every point of Π-$\mathfrak{L}$ has been assigned an ordered pair of elements of $\mathfrak{R}$, and every point of $\mathfrak{L}$-$\{Y\}$ has been assigned a single element of $\mathfrak{R}$ (see Fig. 4.2.)

The lines of Π remain to be named, and shall be named as follows: The line $[(m) \cdot (0, k)]$ will be called $[m, k]$, the line $[Y \cdot (k, 0)]$ will be called

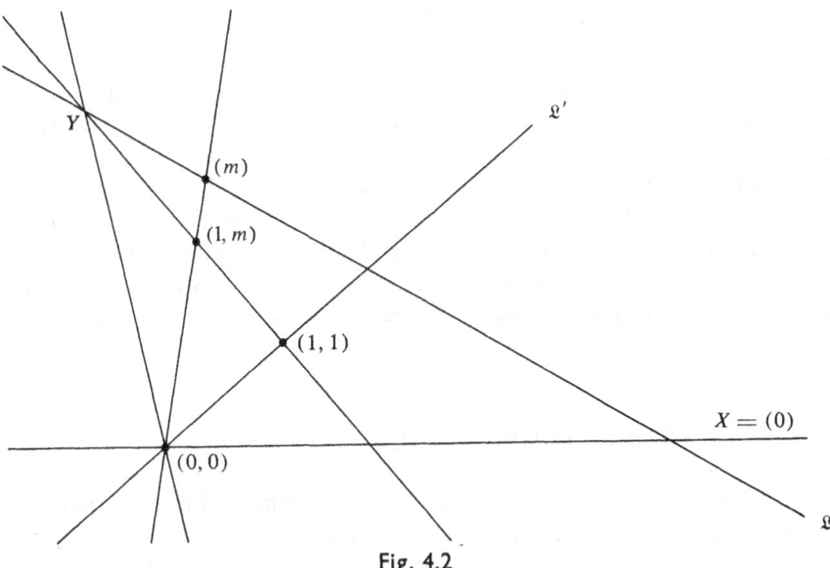

Fig. 4.2

$[k]$, and the line $\mathfrak{L}$, which has not yet been named, will continue to be called "$\mathfrak{L}$." Thus, every line of Π not containing Y has been named by an ordered pair of elements of $\mathfrak{R}$, and every line containing Y, except for $\mathfrak{L}$, has been named by a single element of $\mathfrak{R}$ (see Fig. 4.3).

Exercise

2. Verify that the lines not containing Y are in one-one correspondence with ordered pairs of elements of $\mathfrak{R}$.

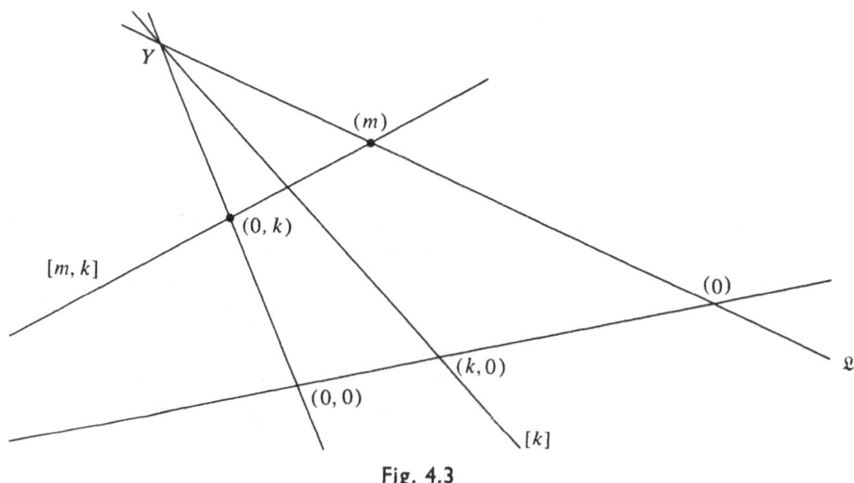

Fig. 4.3

We have now named every point and line of Π using the set $\mathfrak{R}$. We turn next to the problem of introducing an *algebraic* structure determined by the geometry into the set $\mathfrak{R}$, and then of studying that structure and elucidating the nature of the plane. The reader at this point would be well advised to return to Chapter 1 and compare the naming of the elements of an arbitrary plane Π with the discussion of the plane $\Pi_{\mathfrak{R}}$ there. In fact, it will be useful if the reader will keep that example and the discussion of Section 3.5 in mind when studying the general case, and note the analogies as they become apparent.

2. The Planar Ternary Ring

Before actually defining the planar ternary ring, a brief discussion of ternary operations is necessary. In our definition of loops and groups, we introduced the idea of a binary operation on a set, that is, a function from *ordered* pairs of elements of the set into the set. Then, to define loops, we said that every set and binary operation satisfying certain conditions was to be called a loop.

For the description of the algebraic systems which can be obtained from projective planes, we need the idea of a *ternary* function, which is a simple generalization of the binary functions we have already discussed.

Let $\mathfrak{R}$ be a set. If f is a rule that associates with each *ordered triple* of elements of $\mathfrak{R}$, (x_1, x_2, x_3), (where the x_i are in $\mathfrak{R}$), a unique element of $\mathfrak{R}$, then f is called a *ternary operation*:

$$f(x_1, x_2, x_3) = x \in \mathfrak{R}. \tag{1}$$

Now, let Π be a finite projective plane, and let $\mathfrak{R}$ be the set that we have used in the previous section to name the elements of Π. We shall define a ternary operation on $\mathfrak{R}$ using the geometric properties of Π, and then deduce some of its algebraic characteristics from the geometry of Π. Let x, y, m, k be elements of $\mathfrak{R}$. Recall that (x, y) and $[m, k]$ denote, respectively, a particular point and line of Π, and define:

$$y = F(x, m, k) \Leftrightarrow (x, y)\mathrm{I}[m, k]. \tag{2}$$

To begin with, we prove the following lemma.

LEMMA 1. *F is a ternary operation on $\mathfrak{R}$.*

Proof. We need simply prove that to each ordered triple x, m, k, there is a unique element y such that $y = F(x, m, k)$. But from the definition of F, this is equivalent to showing that, for each line $[m, k]$, and each element x, there is a *unique* point (x, y) that is incident with $[m, k]$. By Exercise 1, we know that the line $[Y \cdot (x, x)]$ consists of *all* points of the form (x, a), where a is in $\mathfrak{R}$, plus the point Y. Also, we know that $[m, k]$ is a line that intersects $\mathfrak{L}$ in (m), therefore not in Y, so we can conclude that $[m, k]$ is not incident with Y (Why?). Thus, $[m, k] \cap [Y \cdot (x, x)] = (x, y)$, for some unique y in $\mathfrak{R}$ proving the lemma.

Having established the fact that F is a ternary operation, we would like to explore some of the properties forced on F by its definition and the geometrical configurations on Π. The next theorem shows that there is much that can be said about F.

THEOREM 1. *F satisfies the following conditions:*

(a) $F(a, 0, c) = F(0, b, c) = 0$, for all a, b, c in $\mathfrak{R}$;

(b) $F(1, a, 0) = F(a, 1, 0) = a$, for all a in $\mathfrak{R}$;

(c) Given any a, b, c, d in $\mathfrak{R}$ such that $a \neq c$, there exists a unique x in $\mathfrak{R}$ such that $F(x, a, b) = F(x, c, d)$;

(d) For any a, b, c in $\mathfrak{R}$, there exists a unique x in $\mathfrak{R}$ such that $F(a, b, x) = c$;

(e) For any a, b, c, d in $\mathfrak{R}$ such that $a \neq c$, there exists a unique pair x, y, such that $F(a, x, y) = b$, $F(c, x, y) = d$.

Proof. Part of the proof will be given here, and the rest left as an exercise. For statement (a), for example, we know that $x = F(a, 0, c) \Leftrightarrow (a, x)\,\mathrm{I}\,[0, c]$. But $(a, c)\,\mathrm{I}\,[0, c]$ by Exercise 1. Thus $x = c$, and $F(a, 0, c) = c$. Similarly, $F(0, b, c) = x \Leftrightarrow (0, x)\,\mathrm{I}\,[b, c]$. But $(0, c)\,\mathrm{I}\,[b, c]$ (since $[b, c] = [(b) \cdot (0, c)]$), so $x = c$, and $F(0, b, c) = c$. The other statements follow similarly from the method of coordinatizing Π and the geometric properties satisfied by Π.

Exercise

3. Prove parts (b) to (e) of Theorem 1.

DEFINITION. We shall call a pair $(\mathfrak{R}, F)$, where $\mathfrak{R}$ is a coordinatizing set of a plane, and F its associated ternary operation, *a planar ternary ring.*

Thus, Theorem 1 is a theorem that gives certain properties satisfied by any planar ternary ring. A significant fact about the theorem is that its converse is also true, so that its conditions actually characterize planar ternary rings. We state this as Theorem 2.

THEOREM 2. Let $\mathfrak{R}$ be a set and F a ternary operation on $\mathfrak{R}$ satisfying conditions (a) to (e) of Theorem 1. Then $(\mathfrak{R}, F)$ is a planar ternary ring, that is, there is a plane Π such that $(\mathfrak{R}, F)$ is a planar ternary ring of Π.

Proof. We shall first describe all the points and lines of Π, and then the incidences. The points will be of three types: type (1) will consist of all ordered pairs (x, y) of elements of $\mathfrak{R}$; type (2) will consist of all elements of $\mathfrak{R}$, (x); and type (3) consists of a single "special" point, called Y. Similarly, the lines of Π fall into three classes; type (1) is all ordered pairs of elements of $\mathfrak{R}$, $[x, y]$; type (2) consists of all elements of $\mathfrak{R}$, $[k]$; and type (3), the single line $\mathfrak{L}$. The incidences are defined as follows:

$$
\begin{array}{ll}
(x, y)\,\mathrm{I}\,[m, k] \Leftrightarrow y = F(x, m, k) & \text{for all } x, m, y, k \text{ in } \mathfrak{R}, \\
(x, y)\,\mathrm{I}\,[x] & \text{all } x, y \text{ in } \mathfrak{R}, \\
(m)\,\mathrm{I}\,[m, k] & \text{all } m, k \text{ in } \mathfrak{R}, \\
(m)\,\mathrm{I}\,\mathfrak{L} & \text{all } m \text{ in } \mathfrak{R}, \\
Y\,\mathrm{I}\,\mathfrak{L} & \\
Y\,\mathrm{I}\,[k] & \text{all } k \text{ in } \mathfrak{R}.
\end{array}
$$

The remainder of the proof is left as an exercise.

Exercise

4. Show, by case analyses, that Π, as defined above, does indeed satisfy conditions (a′), (b′), and (c′), of Section 1.3, and is therefore a projective plane when $(\mathfrak{R}, F)$ has the hypothesized properties. (For example, to show that (a′) is satisfied, you must distinguish when two points P and Q are of types (1), (2), (3)—five cases—when checking that they determine a unique line. Thus, if P and Q are both of type (1), then condition (e) of Theorem 2 provides a proof.) Then show that $(\mathfrak{R}, F)$ is the planar ternary ring obtained from Π when the obvious coordinatizing quadrangle $\{(0), Y, (0, 0), (1, 1)\}$ is selected.

A result that will be useful later is the following theorem. It could have been embodied in Theorems 1 and 2, but was left as a separate result so that Theorems 1 and 2 could be stated in full generality. Unlike most of our other results, the following theorem is not necessarily true in general but only holds for the case where $\mathfrak{R}$ is finite.

THEOREM 3. Let $\mathfrak{R}$ be a finite set with ternary operation F, containing elements 0, 1, and satisfying conditions (a), (b), and (d) of Theorem 1. Then $\mathfrak{R}$ satisfies condition (c) if and only if it satisfies condition (e).

Proof. Assume $\mathfrak{R}$ satisfies (c), a consequence of the assumption that two lines determine a unique point. Then for each of the n^2 ordered pairs (x, y), where x and y are in $\mathfrak{R}$, there is an ordered pair (α, β), where $\alpha = F(a, c, y)$ and $\beta = F(c, x, y)$. If the n^2 ordered pairs (α, β) so obtained are distinct, then, since there are only n^2 ordered pairs of elements of $\mathfrak{R}$, each one must be obtained exactly once, including the pair (b, d). Thus, if (e) is not satisfied (that is, if not every pair of points determines a unique line), some pair (α, β) must occur twice. Assume

$$\begin{aligned} F(a, x, y) &= F(a, u, v) = \alpha \\ F(c, x, y) &= F(c, u, v) = \beta \end{aligned} \tag{3}$$

where $(x, y) \neq (u, v)$. If $x = u$, $y \neq v$, then we have $F(a, x, y) = F(a, x, v) = \alpha$. But this contradicts condition (d). Thus $x \neq u$. But then (3) contradicts condition (c). Thus each pair must occur only once, and we can conclude that condition (e) is satisfied. The other implication is left as an exercise.

Exercise

5. Complete the proof of Theorem 3 by showing that, if (e) holds, then (c) holds. (Hint: Define a function f on $\mathfrak{R}$ by: $F(x, a, b) = F(x, c, f(x))$. Show that f is one-one, and hence, since $\mathfrak{R}$ is finite, there is an x in $\mathfrak{R}$ such that $f(x) = d$.)

3. Further Properties of $(\mathfrak{R}, F)$

In this section, we shall explore further some of the algebraic properties imposed on $(\mathfrak{R}, F)$ by the geometry of Π. One of the most interesting results has to do with the definition of two binary operations out of the ternary operation F, and the appearance of their surprising structures.

For any a and b in $\mathfrak{R}$, we define the two operations "+" and "·" by

$$\begin{aligned} a + b &= F(a, 1, b) \\ a \cdot b &= F(a, b, 0). \end{aligned} \tag{4}$$

It is easy to see that this definition gives rise to two *binary* operations (for $x + y$ and $x \cdot y$ are unique) on $\mathfrak{R}$. The surprising fact about these operations, however, is contained in the following theorem.

THEOREM 4. The set $\mathfrak{R}$, together with the operation $+$, written $(\mathfrak{R}, +)$, is a loop with identity $\{0\}$. The set $\mathfrak{R} - \{0\}$, together with the operation $\cdot$, $(\mathfrak{R} - \{0\}, \cdot)$, is a loop with identity $\{1\}$. Also, $x \cdot 0 = 0 \cdot x = 0$ for any x in $\mathfrak{R}$.

Proof. To prove that $(\mathfrak{R}, +)$ is a loop whose identity is $\{0\}$, we begin by showing that

$$a + 0 = 0 + a = a \tag{5}$$

for all a in $\mathfrak{R}$. But, using the definition of "+," (5) is equivalent to $F(a, 1, 0) = F(0, 1, a) = a$ for all a in $\mathfrak{R}$, which is a consequence of conditions (a) and (b) of Theorem 1. Next, we show that the equations

$$a + x = b, \qquad y + a = b \tag{6}$$

can always be uniquely solved for x and y. But $a + x = b$ is equivalent to $F(a, 1, x) = b$, which is uniquely solvable for x by condition (d) of Theorem 1. Also, $y + a = b$ is equivalent to $F(y, 1, a) = b$. But by condition (c) of Theorem 1, $F(y, 1, a) = F(y, 0, b)$ has a unique solution for y, and since $F(y, 0, b) = b$, this solution is the desired one.

Exercises

6. Prove that $(\mathfrak{R} - \{0\}, \cdot)$ is a loop with identity 1.

7. Show that $0 \cdot x = x \cdot 0 = 0$ for all x in $\mathfrak{R}$.

This section concludes with the following definition, whose significance will soon be apparent.

DEFINITION. The planar ternary ring $(\mathfrak{R}, F)$ is said to be *linear* if

$$F(x, m, k) = (x \cdot m) + k \tag{7}$$

for all x, m, k in $\mathfrak{R}$.

The reader should notice that the line $[m, k]$ consists of the point (m) and all points (x, y), with $y = F(x, m, k)$. In the case that $(\mathfrak{R}, F)$ is linear, these points have the form $(x, x \cdot m + k)$, and the analogy between this case, the Euclidean Plane, and the field planes without the special line, where lines also consist of points $(x, xm + k)$, should be clear.

Exercise

8. Show that the finite field $GF(p^n)$ satisfies all the conditions of Theorem 1, where the ternary operation is linear and defined by the two binary operations ($+$ and $\cdot$) in $GF(p^n)$, and where the unit "1" of Theorem 1 is the multiplicative identity "e" of $GF(p^n)$.

4. Collineations and Ternary Rings

One question that is of great interest in the subject of projective planes concerns the relationships between the various ternary rings which can be defined by the same plane. Recall (see Section 1) that the ternary ring was determined by our *choice* of the *ordered set of four points* making up the coordinatizing quadrangle. (Of course, the elements of $\mathfrak{R}$ except for 0, 1 could have different *names*, but except for that difference, the choice of the coordinatizing quadrangle determined the ternary function in terms of the geometry of the plane *relative to that quadrangle*.) Thus, by selecting another ordered set of four points as the coordinate quadrangle, a completely different ternary ring could be obtained. In its most general form, this problem has not yet been solved, that is, given any two planar ternary rings (with the same number of elements) to determine whether they coordinatize the same projective plane. Some progress has been made, however, and the theorem in this section will provide a partial answer to the problem.

To begin with, we need some definitions. Let $(\mathfrak{R}, F)$ and $(\mathfrak{S}, G)$ be two planar ternary rings, and let α be a one-one function of $\mathfrak{R}$ onto $\mathfrak{S}$ (we use the notation $x \rightarrow x^\alpha = y \in \mathfrak{S}$). Then α will be called an *isomorphism* of $(\mathfrak{R}, F)$ with $(\mathfrak{S}, G)$ if

$$\big(F(x, m, k)\big)^\alpha = G(x^\alpha, m^\alpha, k^\alpha), \qquad x, m, k \in \mathfrak{R}. \tag{8}$$

If such an isomorphism exists, $(\mathfrak{R}, F)$ and $(\mathfrak{S}, G)$ will be called *isomorphic*. This notion of isomorphism between two ternary rings is analogous to that of isomorphism between two loops, the only difference being whether the function (loop composition on F) is a binary or a ternary function.

As a partial solution to the problem of the relationship between the ternary rings coordinatizing a plane, we have the following theorem.

THEOREM 5. Let Π be a projective plane, and let $(\mathfrak{R}, F)$ and $(\mathfrak{S}, G)$ be the ternary rings for Π determined by the two coordinatizing quadrangles $\{X, Y, Q, I\}$ and $\{X', Y', Q', I'\}$, respectively. Then $(\mathfrak{R}, F)$ and $(\mathfrak{S}, G)$ are isomorphic if and only if there is a collineation, f, of Π satisfying

$$\begin{aligned} f(X) &= X' \qquad f(Q) = Q' \\ f(Y) &= Y' \qquad f(I) = I'. \end{aligned} \tag{9}$$

Proof. Assume first the existence of such a collineation. Then $f[(Q \cdot I)] = [Q' \cdot I']$, and f gives a one-one mapping of the points of the line $[Q \cdot I]$ onto the points of the line $[Q' \cdot I']$, while $f([Q \cdot I] \cap [X \cdot Y]) = [Q' \cdot I'] \cap [X' \cdot Y']$, so we can define a one-one mapping of the elements of $\mathfrak{R}$ onto the elements of $\mathfrak{S}$ by

$$f(a, a) = (a^\alpha, a^\alpha)', \qquad a \in \mathfrak{R},\ a^\alpha \in \mathfrak{S}. \tag{10}$$

Notice that $0^\alpha = 0$, $1^\alpha = 1$. Also, since f is a collineation, we can show that

$$f(a, b) = (a^\alpha, b^\alpha)' \tag{11}$$

for any a, b in $\mathfrak{R}$ (Why?). Finally, since $f(1, m) = (1, m^\alpha)'$, we have $f(m) = (m^\alpha)'$ (Why?). Now we can also check and see that

$$f[m, k] = [m^\alpha, k^\alpha]', \tag{12}$$

for $f[m, k] = f[(m) \cdot (0, k)] = [(m^\alpha)' \cdot (0, k^\alpha)] = [m^\alpha, k^\alpha]'$. Thus we can write

$$\begin{aligned} F(x, m, k) = y &\Leftrightarrow (x, y)\,\mathrm{I}\,[m, k] \Leftrightarrow f(x, y)\,\mathrm{I}\,f[m, k] \\ &\Leftrightarrow (x^\alpha, y^\alpha)'\,\mathrm{I}\,[m^\alpha, k^\alpha]' \\ &\Leftrightarrow G(x^\alpha, m^\alpha, k^\alpha) = y^\alpha, \end{aligned} \tag{13}$$

and thus α is an isomorphism of $(\mathfrak{R}, F)$ and $(\mathfrak{S}, G)$. To prove the other half of the theorem, assume α is an isomorphism of $(\mathfrak{R}, F)$ and $(\mathfrak{S}, G)$. Then define

$$\begin{aligned} f(a, b) &= (a^\alpha, b^\alpha)' \\ f(m) &= (m^\alpha)' \\ f(Y) &= Y' \\ f[m, k] &= [m^\alpha, k^\alpha]' \\ f[k] &= [k^\alpha]' \\ f[\mathfrak{L}] &= \mathfrak{L}'. \end{aligned} \tag{14}$$

Using (13), it is easy to show that (14) defines a collineation of Π which satisfies (9).

Exercise

9. In the proof of Theorem 5, verify the derivation of (11), and the fact that $f(m) = (m^{\alpha})'$. Also, show that (14) defines a collineation of Π.

COROLLARY. Let $\Pi = PG(2,p^n)$. Then any two planar ternary rings of Π are isomorphic.

Proof. This corollary is a direct consequence of Theorems 3.3 and 4.5. We conclude this chapter with a simple consequence of Theorem 2 above.

THEOREM. Let $(\mathfrak{R}, F)$ and $(\mathfrak{S}, G)$ be two isomorphic planar ternary rings. Then they coordinatize isomorphic projective planes.

Proof. Define the planes as in the proof of Theorem 2. Then if α is the isomorphism from $(\mathfrak{R}, F)$ to $(\mathfrak{S}, G)$ we can use the methods of the proof of Theorem 5 to construct an isomorphism between the two planes by using α to define this isomorphism.

Exercise

10. Complete the proof of Theorem 6.

[5]

Central Collineations and the Little Desargues' Property

1. Central Collineations

In Chapter 5, we shall discuss geometric conditions similar to Desargues' configuration (see Section 1.8) and study the implications such conditions have for the planes that satisfy them. To begin with, however, we must introduce and study a certain type of collineation called a "central collineation." This concept has already arisen in Theorem 2.6, and is extremely important, inasmuch as it permits us to study these restricted Desarguesian conditions.

A $(P, \mathfrak{L})$-*central collineation*, f, is a collineation with the property that every point of the line $\mathfrak{L}$ is a fixed point under f and every line containing the point P is a fixed line under f. The point P and the line $\mathfrak{L}$ are called the *center* and *axis* of f, respectively. If $P \mathrel{I} \mathfrak{L}$, then f is called a *translation*; if $P \not{I} \mathfrak{L}$, f is called a *homology*. A $(P, \mathfrak{L})$-central collineation will simply be called a *central collineation* if we do not wish to specify the axis and center. Observe that the identity collineation is a $(P, \mathfrak{L})$-central collineation for every point P and line $\mathfrak{L}$ of Π.

We begin our study of central collineations with Lemma 1.

LEMMA 1. Let f be a collineation of Π, not its identity collineation, i, and let f fix all of the points of some line $\mathfrak{L}$. Then there is at most one point of Π fixed by f that is not on $\mathfrak{L}$.

Proof. Let P and Q be two fixed points under f which are not on $\mathfrak{L}$. Then if R is any point not on $[P \cdot Q]$ or $\mathfrak{L}$, we can write $R = [R \cdot P] \cap [R \cdot Q]$ But any line through P or Q is a fixed line (Why?), so R is the intersection of two fixed lines and hence a fixed point. If $R \mathrel{I} [P \cdot Q]$, choose R_1 not on $[P \cdot Q]$. Then, by the above argument, R_1 must be fixed, and we can show that R must also be fixed, since $R \not{I} [R_1 \cdot P]$. Thus every point of Π is fixed, and $f = i$.

LEMMA 2. Let $f \neq i$ be as in Lemma 1. Then there is a point P such that all lines containing P are fixed.

Proof. If there is a fixed point $P \not{I} \mathfrak{L}$, then every line containing P must be fixed (Why?). If all the fixed points are on $\mathfrak{L}$, let Q be any point not on $\mathfrak{L}$, and define the point P by $P = [Q \cdot f(Q)] \cap \mathfrak{L}$. Clearly, the line $[Q \cdot f(Q)]$ is fixed (Why?). In fact, for any point $Q_1 \not{I} \mathfrak{L}$, $[Q_1 \cdot f(Q_1)]$ is fixed. Thus, $[Q \cdot f(Q)] \cap [Q_1 \cdot f(Q_1)]$ is fixed, and therefore must be on $\mathfrak{L}$, and hence is equal to P. Thus, for *any* $Q_1 \not{I} \mathfrak{L}$, $[Q_1 \cdot f(Q_1)] = [P \cdot Q_1]$ is a fixed line so *every* line containing P is fixed, proving the lemma. Using the lemmas, we can prove

THEOREM 1. Let f be a collineation fixing every point on some line $\mathfrak{L}$. Then f is a central collineation with $\mathfrak{L}$ as axis. If $f \neq i$, the identity, then its axis and center are unique.

Proof. All but the last sentence is a direct corollary of the previous two lemmas. Lemma 1 shows that the axis is unique, and the dual of Lemma 1, that the center is unique.

Exercises

1. Fill in the details of the proof of Theorem 1.

2. If $f \circ f = i$ and $f \neq i$, and if f is a central collineation (see Theorem 2.6), then show that if the order of the plane is even, f is a translation, and if the order is odd, then f is a homology.

Now let Π be any projective plane and, as before, let G_Π be the group of all collineations of Π. Then, for any point P and line $\mathfrak{L}$, we *define* the following sets:

$G_{P,\mathfrak{L}} = \{$all $(P, \mathfrak{L})$-central collineations, including $i\}$;

$G_{P,-} = \{$all central collineations with P as center, including $i\}$;

$G_{-,\mathfrak{L}} = \{$all central collineations with $\mathfrak{L}$ as axis, including $i\}$.

Using Lemmas 1 and 2 and Theorem 1, we can easily prove the following lemma.

LEMMA 3. The three sets defined above all form subgroups of the group G_Π.

Proof. We need only prove that each of the sets is closed under composition and taking inverses, that is, if f is in $G_{P,\mathfrak{L}}$ and g is in $G_{P,\mathfrak{L}}$ then $f \circ g$ is in $G_{P,\mathfrak{L}}$, and f^{-1} is in $G_{P,\mathfrak{L}}$. But this follows easily from the uniqueness of the center and axis asserted by Theorem 1.

Exercise

3. Complete the proof of Lemma 3.

We can now define a notion that will prove to be central to the remainder of our study: The projective plane Π is said to be $(P, \mathfrak{L})$-*transitive* if, for any pair of points $Q_1 \neq P$, $Q_2 \neq P$, $Q_1 \not{I} \mathfrak{L}$, $Q_2 \not{I} \mathfrak{L}$, and $P\, I\, [Q_1 \cdot Q_2]$, there exists a collineation f in $G_{P,\mathfrak{L}}$ such that $f(Q_1) = Q_2$ (see Fig. 5.1).

We can make the concept of $(P, \mathfrak{L})$-transitivity easier to apply by proving Theorem 2.

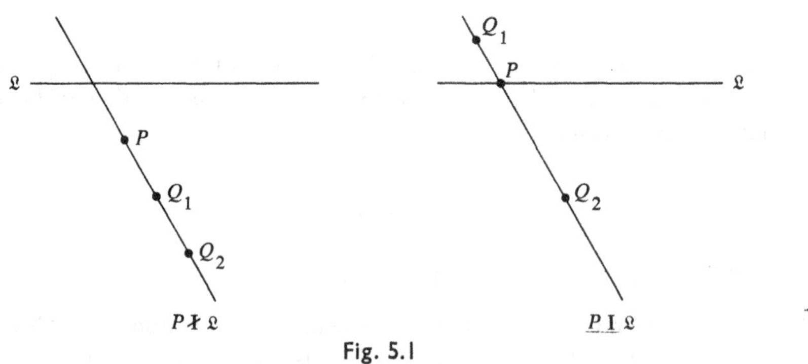

Fig. 5.1

THEOREM 2. If there is a point $Q \neq P$, $Q \not{I} \mathfrak{L}$, such that for every point $R\, I\, [P \cdot Q]$, $R \neq P$, $R \not{I} \mathfrak{L}$, there is a collineation f of $G_{P,\mathfrak{L}}$ such that $f(Q) = R$, then Π is $(P, \mathfrak{L})$-transitive.

Before we prove this theorem, let us observe how much simpler it is to show that for any R *on one line* $[P \cdot Q]$ we can find the desired collineation, than to try to show that for *every pair of points* collinear with P there exists the right kind of collineation. In the first case, we have of the order of n collineations to check, and in the second, we must seemingly show the existence of roughly n^2 collineations.

Proof. First, let R_1 and R_2 be any two points, distinct from P and not incident with $\mathfrak{L}$, on the line $[P \cdot Q]$. Then, if f is the element of $G_{P,\mathfrak{L}}$ satisfying $f(Q) = R_1$, and g in $G_{P,\mathfrak{L}}$ satisfies $g(Q) = R_2$, then $g \circ f^{-1}(R_1) = R_2$ and $g \circ f^{-1}$ is in $G_{P,\mathfrak{L}}$, by Lemma 3. Now let Q_1 and Q_2 be any two points distinct from P and not incident with $\mathfrak{L}$ which are collinear with P.

Then we wish to show that there exists an element f in $G_{P,\mathfrak{L}}$ such that $f(Q_1) = Q_2$. But let R be any point of $\mathfrak{L}$ *except* $\mathfrak{L} \cap [Q \cdot P]$ or $\mathfrak{L} \cap [Q_1 \cdot Q_2]$. Then (see Fig. 5.2) if we define $R_1 = [R \cdot Q_1] \cap [Q \cdot P]$, and $R_2 = [R \cdot Q_2] \cap [Q \cdot P]$, we know by the first part of the proof that there is a collineation f in $G_{P,\mathfrak{L}}$ such that $f(R_1) = R_2$. But let us now make use of the fact that f is in $G_{P,\mathfrak{L}}$, so $f(R) = R$, and $f[Q \cdot P] = [Q \cdot P]$. Also, $f[Q_1 \cdot Q_2] = [Q_1 \cdot Q_2]$. Now, $Q_1 = [R \cdot R_1] \cap [Q_1 \cdot Q_2]$, while $Q_2 = [R \cdot R_2] \cap [Q_1 \cdot Q_2]$. But $f(Q_1) = f[R \cdot R_1] \cap f[Q_1 \cdot Q_2] = [R \cdot R_2] \cap [Q_1 \cdot Q_2]$, so f is the desired collineation, and we have proved the theorem.

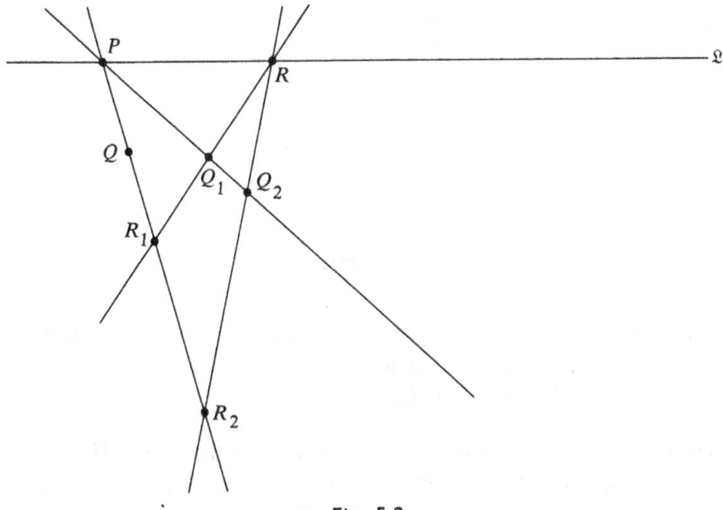

Fig. 5.2

To make immediate use of Theorem 2, let us state and prove the following important theorem.

THEOREM 3. Let Π be $(P, \mathfrak{L})$-transitive and $(Q, \mathfrak{L})$-transitive, where $P \mathrm{I} \mathfrak{L}$, $Q \mathrm{I} \mathfrak{L}$, and $P \neq Q$. Then Π is $(R, \mathfrak{L})$-transitive for every $R \mathrm{I} \mathfrak{L}$.

Proof. The idea of the proof is expressed in Fig. 5.3. Let R be any point of $\mathfrak{L}$ except P or Q, and let P_1 and P_2 be any two points of Π (not on $\mathfrak{L}$) which are collinear with R. Then we construct the lines $\mathfrak{L}_1 = [P_1 \cdot P]$ and $\mathfrak{L}_2 = [P_2 \cdot Q]$, and, finally, let $Q_1 = (\mathfrak{L}_1 \cap \mathfrak{L}_2)$. Since Π is $(P, \mathfrak{L})$-transitive, there is an element f of $G_{P,\mathfrak{L}}$ such that $f(P_1) = Q_1$. Also, since Π is $(Q, \mathfrak{L})$-transitive, there is a g in $G_{Q,\mathfrak{L}}$ with $g(Q_1) = P_2$. Thus, $g \circ f(P_1) = P_2$, and $g \circ f$ is in $G_{-,\mathfrak{L}}$ (Why?). Now, if we can show that the center of $g \circ f$ is R, we shall be finished. But since $[P_1 \cdot P_2] = [P \cdot P_1] = [P \cdot P_2]$ is a fixed line (Why?), the center of $g \circ f$ must be on $[P_1 \cdot P_2]$. The remainder of the proof is left as an exercise.

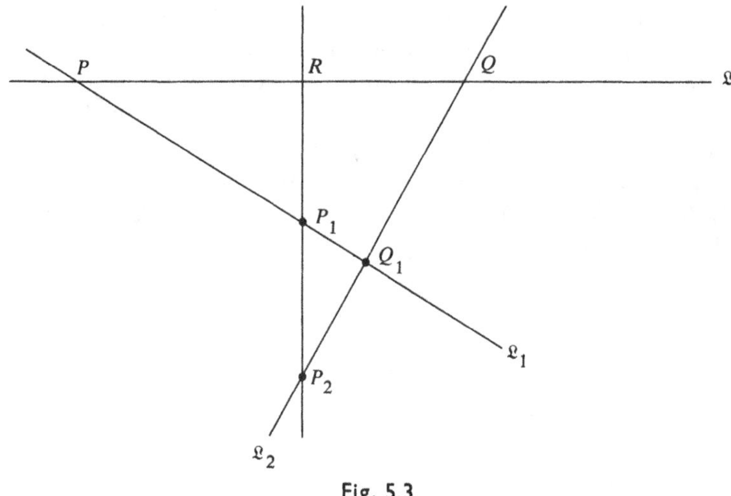

Fig. 5.3

Exercise

4. Complete the proof of Theorem 3 by showing that the center of $g \circ f$ must be on $\mathfrak{L}$, and hence is equal to $\mathfrak{L} \cap [P_1 \cdot P_2] = R$. (Hint: Let Q_3 be any point of $[P_1 \cdot P_2]$ other than R, and show that $g \circ f(Q_3) \neq Q_3$.)

Theorem 3 leads us to make the following definition: If Π is $(P, \mathfrak{L})$-transitive for *every* $P \mathrm{I} \mathfrak{L}$, we shall say that Π is $(\mathfrak{L}, \mathfrak{L})$-*transitive.* Also, if Π is $(P, \mathfrak{L})$-transitive for every $\mathfrak{L} \mathrm{I} P$, we shall say that Π is (P, P)-*transitive.* We shall see in subsequent sections the strong consequences of $(\mathfrak{L}, \mathfrak{L})$-transitivity on the algebraic structure of the coordinatizing planar ternary rings.

2. Little Desargues' Property

In this section, we shall define and study a concept similar to the notion we defined in Section 1.8, and we shall use the notation of that section with regard to triangles in a projective plane. Let $\{A_1, A_2, A_3\}$, and $\{B_1, B_2, B_3\}$ be any two triangles in a plane Π that are perspective from a point P. If $\mathfrak{L}$ is a line of Π such that whenever $([A_1 \cdot A_2] \cap [B_1 \cdot B_2]) \mathrm{I} \mathfrak{L}$ and $([A_1 \cdot A_3] \cap [B_1 \cdot B_3]) \mathrm{I} \mathfrak{L}$, we *must* have $([A_2 \cdot A_3] \cap [B_2 \cdot B_3]) \mathrm{I} \mathfrak{L}$, then we shall say that Π is $(P, \mathfrak{L})$-*Desarguesian.* Thus, Π is Desarguesian if and only if Π is $(P, \mathfrak{L})$-Desarguesian for all pairs $(P, \mathfrak{L})$.

Exercises

5. Prove the last assertion.

6. Show that Π is $(P, \mathfrak{L})$-Desarguesian, if and only if given any two triangles $\{A_1, A_2, A_3\}$, $\{B_1, B_2, B_3\}$ perspective from $\mathfrak{L}$ and such that $P \mathrm{I} [A_1 \cdot B_1]$ and $P \mathrm{I} [A_2 \cdot B_2]$, then we must have $P \mathrm{I} [A_3 \cdot B_3]$.

This restricted kind of Desargues' property is obviously a much weaker condition than the full Desargues' property. It is an extremely useful tool, however, in the study of projective planes, as we can see in the next result which ties together the study of the geometric configurations in a plane and the study of that plane's collineation group.

THEOREM 4. A plane Π is $(P, \mathfrak{L})$-Desarguesian if and only if it is $(P, \mathfrak{L})$-transitive.

Proof. As a first step in the proof, we shall need the following lemma.

LEMMA 4. Let Π be a projective plane, and let f be a central collineation of Π with center P and axis $\mathfrak{L}$. If $\{A_1, A_2, A_3\} = \Delta$ is any triangle, no point of which lies on $\mathfrak{L}$ and no line of which contains P, then $\{f(A_1), f(A_2), f(A_3)\} = \Delta'$ is a triangle such that Δ and Δ' are perspective from both P and $\mathfrak{L}$.

Proof of the Lemma. Since P is the center of f, we know that P, A_i, and $f(A_i)$ are collinear for $i = 1, 2, 3$ (Why?), so the two triangles are perspective from P. Similarly, since every point of $\mathfrak{L}$ is fixed, $([A_1 \cdot A_2] \cap L) = X$ is fixed, and

$$f[A_1 \cdot A_2] = f[A_1 \cdot X] = [X \cdot f(A_1)] = [X \cdot f(A_2)] = [f(A_1) \cdot f(A_2)].$$

Hence $([A_1 \cdot A_2] \cap [f(A_1) \cdot f(A_2)]) = X$, which is incident with $\mathfrak{L}$, and we can conclude that Δ and Δ' are perspective from $\mathfrak{L}$ (Why?).

We can now proceed to the proof of Theorem 4. Assume first that Π is $(P, \mathfrak{L})$-transitive, and let $\{A_1, A_2, A_3\}$ and $\{B_1, B_2, B_3\}$ be any two triangles perspective from P and such that $X = ([A_1 \cdot A_2] \cap [B_1 \cdot B_2])$ is incident with $\mathfrak{L}$ and $Y = ([A_1 \cdot A_3] \cap [B_1 \cdot B_3])$ is incident with $\mathfrak{L}$. We wish to prove that $([A_2 \cdot A_3] \cap [B_2 \cdot B_3]) \mathrm{I} \mathfrak{L}$. To that end, let f be the $(P, \mathfrak{L})$-central collineation with the property $f(A_1) = B_1$. Since f has center P and axis $\mathfrak{L}$, we can deduce that

$$\begin{aligned} f(A_3) &= f([A_1 \cdot Y] \cap [A_3 \cdot B_3]) = f[A_1 \cdot Y] \cap f[A_3 \cdot B_3] \\ &= [B_1 \cdot Y] \cap [A_3 \cdot B_3] = B_3. \end{aligned} \tag{1}$$

Next, we can prove that $f(A_2) = B_2$ by a similar argument, and Lemma 4 now completes this part of the argument.

Now let us assume that Π is $(P, \mathfrak{L})$-Desarguesian and try to show that it must be $(P, \mathfrak{L})$-transitive. To begin the proof, let X, Y be two points collinear with P, but not on $\mathfrak{L}$. We wish to *define* a $(P, \mathfrak{L})$-central collineation, f, with the property

$$f(X) = Y. \tag{2}$$

Thus $f(P) = P$, and $f(Q) = Q$ for any point Q on $\mathfrak{L}$. To define f on $\Pi - \{\mathfrak{L}\}$, we begin by letting R be any point not on $\mathfrak{L}$, not on $[X \cdot Y]$, and not P. Now, let

$$R_1 = ([X \cdot R] \cap \mathfrak{L}). \tag{3}$$

Then, $R = [R_1 \cdot X] \cap [P \cdot R]$. Since f is to be a $(P, \mathfrak{L})$-central collineation, we *must* define

$$f(R) = f[R_1 \cdot X] \cap f[P \cdot R] = [R_1 \cdot Y] \cap [P \cdot R]. \tag{4}$$

In this way, we can define f on all points of Π except the points other than P and X of $[X \cdot Y]$. In order to define f on $[X \cdot Y]$, we can use the definition already made on the points of some other line through P, and extend f according to the rules of (4). To demonstrate that f is as desired, we must now prove two facts: (i) the definition of f on $\Pi - \{[X \cdot Y]\}$ is such that collinear points are taken into collinear points by f, and (ii) no matter which pair of points $(Q, f(Q))$ we use to extend f to the points of $[X \cdot Y]$, the result is the same. The second fact proves that f is well defined by our method, and (i) shows that f can be extended to the lines of Π in the obvious way (that is, if $Q_1 \mathrm{I} \mathfrak{M}$ and $Q_2 \mathrm{I} \mathfrak{M}$, define $f(\mathfrak{M}) = [f(Q_1) \cdot f(Q_2)]$). The proofs of (i) and (ii) are similar, and make use of the fact that Π is assumed $(P, \mathfrak{L})$-Desarguesian. To prove (i), it suffices to show that $f(Q_1), f(Q_2)$, and $([Q_1 \cdot Q_2] \cap \mathfrak{L})$ are collinear for any points Q_1 and Q_2 not on $[X \cdot Y]$ or $\mathfrak{L}$, and not collinear with P. (Why does it suffice?) We may as well assume that X, Q_1 and Q_2 are not collinear, because if they are, $([X \cdot Q_1] \cap \mathfrak{L}) = ([X \cdot Q_2] \cap \mathfrak{L})$, and $f(Q_1)$ and $f(Q_2)$ must be collinear with Y and $([X \cdot Q_1] \cap \mathfrak{L})$. So (see Fig. 5.4) let Q_1 and Q_2 be as stated, and let $X_1 = ([X \cdot Q_1] \cap \mathfrak{L})$, $X_2 = ([X \cdot Q_2] \cap \mathfrak{L})$. Thus, $f(Q_1) = ([X_1 \cdot Y] \cap [P \cdot Q_1])$ and $f(Q_2) = ([X_2 \cdot Y] \cap [P \cdot Q_2])$. Now consider the two triangles $\{X, Q_1, Q_2\}$ and $\{Y, f(Q_1), f(Q_2)\}$. They are perspective from P, and $[X \cdot Q_1] \cap [Y \cdot f(Q_1)$ $= X_1$ which is on $\mathfrak{L}$, while $([X \cdot Q_2] \cap [Y \cdot f(Q_2)]) = X_2$ which is also on $\mathfrak{L}]$ Thus, since Π is $(P, \mathfrak{L})$-Desarguesian, $([Q_1 \cdot Q_2] \cap [f(Q_1) \cdot f(Q_2)]) \mathrm{I} \mathfrak{L}$. But this conclusion proves (i). To prove that (ii) is true, let Q_1, Q_2 be any. two points not on $[X \cdot Y]$, not on $\mathfrak{L}$, and not collinear with P. Then, using (4), the mapping $X \to Y$ has defined points $f(Q_1)$ and $f(Q_2)$. Now if P_1 is any point on $[X \cdot Y]$, let us prove that the pairs $(Q_1, f(Q_1))$ and $(Q_2, f(Q_2))$ imply, by (4), the *same* mapping on P_1 (see Fig. 5.5). First, by (i), we may assume that Q_1, Q_2, and P_1 are not collinear, or the condition is easily seen to

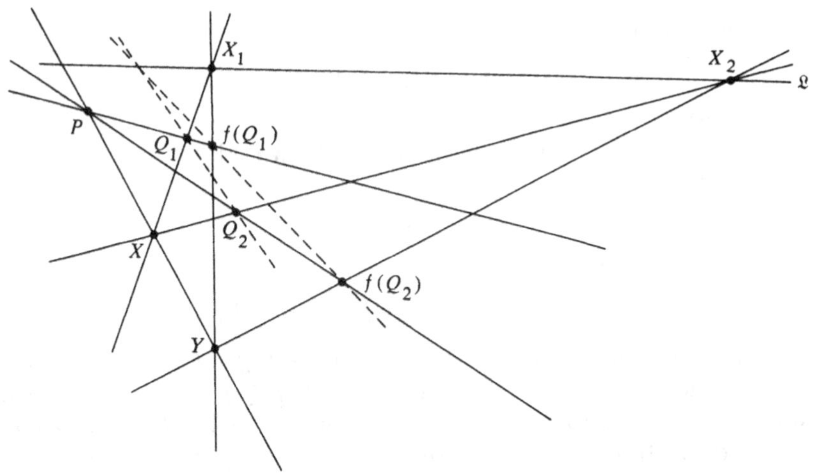
X_1
X_2
$\mathfrak{L}$
P
Q_1
$f(Q_1)$
Q_2
X
$f(Q_2)$
Y

Fig. 5.4

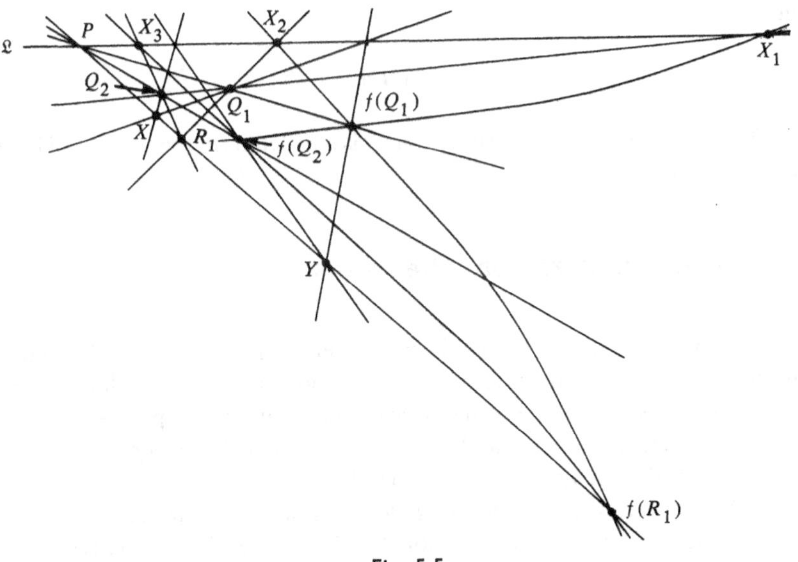
$\mathfrak{L}$
P
X_3
X_2
X_1
Q_2
Q_1
$f(Q_1)$
X
R_1
$f(Q_2)$
Y
$f(R_1)$

Fig. 5.5

be satisfied (Why?). Now define

$$X_1 = [Q_1 \cdot Q_2] \cap [f(Q_1) \cdot f(Q_2)], \tag{5}$$

and note that $X_1 \, \mathrm{I} \, \mathfrak{L}$, by (i). Then define

$$X_2 = [R_1 \cdot Q_1] \cap \mathfrak{L}, \tag{6}$$

and by (4),

$$f(R_1) = [X_2 \cdot f(Q_1)] \cap [X \cdot Y]. \tag{7}$$

Finally, define

$$X_3 = [R_1 \cdot Q_2] \cap \mathfrak{L}, \tag{8}$$

and we must prove that

$$f(R_1) = [X_3 \cdot f(Q_2)] \cap [X \cdot Y]. \tag{9}$$

But the two triangles $\{Q_1, Q_2, R_1\}$ and $\{f(Q_1), f(Q_2), f(R_1)\}$ are perspective from P, and (5), (6), and (7) imply that $[f(R_1) \cdot f(Q_2)] \cap [R_1 \cdot Q_2]$ is incident with $\mathfrak{L}$ (since Π is assumed $(P, \mathfrak{L})$-Desarguesian). But if $([f(R_1) \cdot f(Q_2)] \cap \mathfrak{L}) = X_3$, then (9) must hold, and (ii) is proved, completing the proof of the theorem.

The proof of Theorem 4 is the longest and most complicated we have encountered so far. It is well worth the effort because, in studying projective planes, one of the two conditions is often arrived at and it is most helpful to be able to conclude the truth of the other one. In fact, the result will be found most useful in the proofs of many of the remaining theorems in this chapter.

Exercise

7. Verify Equation (1) and supply the missing details in the proof of Theorem 4, mostly marked "(Why?)."

3. Coordinatization Theorems

We proceed now to the study of the relationship between the algebraic properties of the planar ternary rings coordinatizing a plane and the geometric properties of the plane. Recall that we used four points $\{X, Y, Q, I\}$ to coordinatize a plane Π, and that every point of Π, *except* Y, received a name using the elements of the ring $\mathfrak{R}$. Also, every line of Π, *except* $\mathfrak{L} = [X \cdot Y]$, received a name using elements of $\mathfrak{R}$. *For the remainder of this chapter*, we shall reserve the symbols "Y_S" and "$\mathfrak{L}_S$" to refer, respectively, to this *special* point and line in the coordinatization process: $Y_S = Y$, $\mathfrak{L}_S = \mathfrak{L}$. We do this since this pair of elements plays an important and unique role in all of the remaining theorems of this chapter.

In Chapter 4, we defined the notion of the linearity of a planar ternary ring in terms of the two operations $+$ and $\cdot$ (see Section 4.3). We can now restate the linearity of the ternary ring as a restricted kind of $(Y_S, \mathfrak{L}_S)$-Desarguesian property:

THEOREM 5. Let Π be a projective plane, and $(\mathfrak{R}, F)$ a coordinatizing ring of Π determined by the quadrangle $\{X, Y, Q, I\}$. Then $F(x, m, k) = x \cdot m + k$ if and only if whenever two triangles $\{A_1, A_2, A_3\}$ and $\{B_1, B_2, B_3\}$ satisfy the properties (i) they are perspective from Y_S, (ii) the line $[A_1 \cdot B_1] = [0] = [Y_S \cdot (0, 0)]$, (iii) the points $[A_1 \cdot A_2] \cap [B_1 \cdot B_2]$ and $[A_1 \cdot A_3] \cap [B_1 \cdot B_3]$ are both on $\mathfrak{L}_S$, and (iv) the point $([A_2 \cdot A_3] \cap \mathfrak{L}_S) = (0)$, then (v) the intersection of $[B_2 \cdot B_3]$ and $\mathfrak{L}_S$ is the point (0).

Before proving the theorem, we would like to observe one of its consequences. Notice that when Π is $(Y_S, \mathfrak{L}_S)$-Desarguesian, conditions (i) to (iii) imply that $[A_2 \cdot A_3] \cap [B_2 \cdot B_3]$ is on $\mathfrak{L}_S$, so an easy corollary of Theorem 5 states that *when Π is $(Y_S, \mathfrak{L}_S)$-Desarguesian, $(\mathfrak{R}, F)$ is linear.* The reader should always keep in mind that the ternary ring under discussion is a particular one of the many ternary rings for Π, namely, the one determined by the quadrangle (X, Y, Q, I). It should be noted, however, that many implications hold for *several* coordinate systems. For example, since the theorem makes no mention of Q and I, we can conclude that, *if Π is $(P, \mathfrak{M})$-Desarguesian, then any ternary ring determined by a quadrangle $\{X, Y, Q, I\}$ with $Y = P$ and $X \mathrm{I} \mathfrak{M}$ is linear* (Why?). In general, all of the theorems of this section must be read with the fact in mind that a particular geometric condition often implies a certain algebraic property for *several* ternary rings of the plane.

Let us now prove the theorem: We first take two triangles satisfying (i) to (iv) (see Fig. 5.6) and show that linearity of $(\mathfrak{R}, F)$ holds if and only if (v) is true. Since, by (ii), $A_1 \mathrm{I} [0]$ and $B_1 \mathrm{I} [0]$, we can write $A_1 = (0, a)$ and $B_1 = (0, b)$. Also let $(m) = ([A_1 \cdot A_2] \cap [B_1 \cdot B_2])$ and

$$(n) = ([A_1 \cdot A_3] \cap [B_1 \cdot B_3]).$$

Since A_2, A_3, and (0) are collinear [from (iv)], we can write $A_2 = (u, b)$, $A_3 = (v, b)$, and then [from (i)] we have $B_2 = (u, c)$, $B_3 = (v, f)$. Now, condition (v) is equivalent to the collinearity of the points, B_2, B_3, and (0), that is,

$$c = f. \tag{10}$$

First assume that $(\mathfrak{R}, F)$ is linear. Then, since $A_2 \mathrm{I} [(m) \cdot A_1]$, or $(u, b) \mathrm{I} [m, a]$ and also $(v, b) \mathrm{I} [n, a]$, we can write

$$b = (u \cdot m) + a = (v \cdot n) + a \tag{11}$$

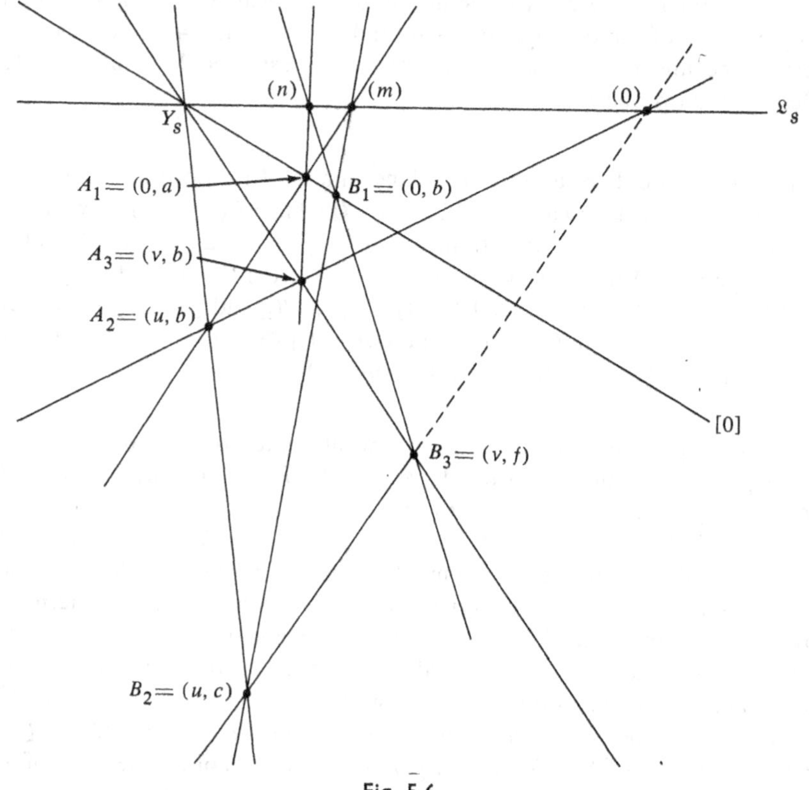

Fig. 5.6

(Why?). Similarly, since $B_2 \mathrm{I} [(m) \cdot B_1]$ and $B_3 \mathrm{I} [(n) \cdot B_1]$, we have

$$c = (u \cdot m) + d \tag{12}$$

and

$$f = (v \cdot n) + d. \tag{13}$$

Since $(\mathfrak{R}, +)$ is a loop, (10) implies that $u \cdot m = v \cdot n$, and using this fact in (12) and (13), we can conclude that $c = f$, proving half of the theorem. Next, assume that $c = f$ for any values of a, d, u, m, v. Then the incidences that hold imply

$$F(u, m, a) = b = F(v, n, a) \tag{14}$$

and

$$F(u, m, d) = c = F(v, n, d). \tag{15}$$

If these equalities hold for *all* values of a, d, u, m, v, they must hold if we replace some of these variables by particular values. If, for example, we set $d = 0$ in (15) we get

$$c = u \cdot m = v \cdot n. \tag{16}$$

Next, if we set $n = 1$, we have $v = u \cdot m$ from (16), while in (14) we have $F(u, m, a) = F(u \cdot m, 1, a) = (u \cdot m) + a$. Since u, m, and a are still unspecified, we have reached the desired conclusion that $(\mathfrak{R}, F)$ is linear when $c = f$, and so have concluded the proof of the theorem.

The next theorem begins to explore more thoroughly the geometric consequences of assuming more algebraic "structure" in $(\mathfrak{R}, F)$.

THEOREM 6. The ternary ring $(\mathfrak{R}, F)$ is linear and $+$ is associative, that is, $(\mathfrak{R}, +)$ is a group, if and only if Π is $(Y_S, \mathfrak{L}_S)$-Desarguesian.

Proof. Assume first that $(\mathfrak{R}, F)$ is linear and has associative addition. We shall show that Π is $(Y_S, \mathfrak{L}_S)$-transitive, and Theorem 4 implies that Π is $(Y_S, \mathfrak{L}_S)$-Desarguesian. To this end, consider the mapping f_a defined for any a in $\mathfrak{R}$ by:

$$\begin{aligned} (x, y) &\to (x, y + a) & [m, k] &\to [m, k + a] \\ (m) &\to (m) & [k] &\to [k] \\ Y_S &\to Y_S & \mathfrak{L}_S &\to \mathfrak{L}_S. \end{aligned} \tag{17}$$

To show that f_a is a collineation for any a, we must check cases. For example: $Y_S \mathrm{I} \mathfrak{L}_S$ and $f_a(Y_S) \mathrm{I} f_a(\mathfrak{L}_S)$, $(m) \mathrm{I} [m, k]$ and $f_a(m) \mathrm{I} f_a[m, k]$. The most difficult case is when $(x, y) \mathrm{I} [m, k]$. Then, by linearity,

$$y = (x \cdot m) + k. \tag{18}$$

But $f_a(x, y) = (x, y + a)$ and $f_a[m, k] = [m, k + a]$. Now,

$$\begin{aligned} &(x, y + a) \mathrm{I} [m, k + a] \\ &\quad \Leftrightarrow y + a = (x \cdot m) + (k + a) = ((x \cdot m) + k) + a = y + a \text{ (using (18))}. \end{aligned}$$

Thus, $(x, y) \mathrm{I} [m, k] \Leftrightarrow f_a(x, y) \mathrm{I} f_a[m, k]$. The other cases are easily checked, and it is seen that f_a is a collineation. But f_a fixes every point on $\mathfrak{L}$ and every line through Y_S, and so f_a is a $(Y_S, \mathfrak{L}_S)$-central collineation satisfying

$$f_a(0, 0) = (0, a). \tag{19}$$

Now, since f_a is a collineation for any a of $\mathfrak{R}$, Theorem 2 immediately implies that Π is $(Y_S, \mathfrak{L}_S)$-transitive, and we have proved half of the theorem. To prove the reverse implication, assume that Π is $(Y_S, \mathfrak{L}_S)$-Desarguesian. Then Π is $(Y_S, \mathfrak{L}_S)$-transitive, and for every a in $\mathfrak{R}$, there is a $(Y_S, \mathfrak{L}_S)$-central collineation f_a satisfying (19). Also, by the result of Theorem 5, we know that $(\mathfrak{R}, F)$ must be linear, so all we have to show is that $(\mathfrak{R}, +)$ is associative. For any four elements x, y, m, k in $\mathfrak{R}$, we have

$$(x, y) \mathrm{I} [m, k] \Leftrightarrow f_a(x, y) \mathrm{I} f_a[m, k] \Leftrightarrow y = (x \cdot m) + k. \tag{20}$$

We can write $f_a(x, y) = (F_a(x, y), G_a(x, y))$, where F_a and G_a are two functions of $\mathfrak{R}$, $\mathfrak{R} \to \mathfrak{R}$, whose properties we shall determine. Since f_a is a $(Y_S, \mathfrak{L}_S)$-central collineation, $F_a(x, y) = x$ (Why?) for any x and y in $\mathfrak{R}$, so $f_a(x, y) = (x, G_a(x, y))$. Next, we shall show that $G_a(x, y)$ is independent of x, that is,

$$G_a(x, y) = G_a(0, y), \tag{21}$$

for all x, y in $\mathfrak{R}$. But $(x, y) = ([x] \cap [0, y]) = ([x] \cap [(0) \cdot (0, y)])$, so

$$\begin{aligned} f_a(x, y) &= (0, G_a(x, y)) = (f_a[x] \cap f_a[(0) \cdot (0, y)]) = ([x] \cap [(0) \cdot f_a(0, y)]) \\ &= ([x] \cap [(0) \cdot (0, G_a(0, y))]) = (x, G_a(0, y)), \end{aligned}$$

proving (21). Now, $f_a[m, k] = [m, G_a(0, k)]$, so we can write, using (20) and (21),

$$\begin{aligned} y = (x \cdot m) + k \Leftrightarrow G_a(0, y) \\ = (x \cdot m) + G_a(0, k) \\ = G_a(0, x \cdot m + k), \end{aligned} \tag{22}$$

for all x, m, k in $\mathfrak{R}$.

Since $f_a(0, 0) = (0, a)$, we know that $G_a(0, 0) = a$. In (22), let $k = 0$, and conclude

$$G_a(0, x \cdot m) = (x \cdot m) + a \tag{23}$$

for all x, m in $\mathfrak{R}$.

In (23), let $m = 1$ and we have

$$G_a(0, x) = x + a \tag{24}$$

for all x, a in $\mathfrak{R}$.

Using (23) in (22), we can write

$$\begin{aligned} y = (x \cdot m) + k \Leftrightarrow (x \cdot m + k) + a \\ = (x \cdot m) + (k + a) \end{aligned} \tag{25}$$

for all m, k, a in $\mathfrak{R}$.

Finally, let $m = 1$ in (25), and we can conclude $(x + k) + a = x + (k + a)$ for all x, k, a in $\mathfrak{R}$, and our proof is complete.

Exercises

8. Complete the case checking required to prove that f_a is a collineation in the proof of the first half of the theorem.

9. Supply a proof of Theorem 6 similar in nature to the proof of Theorem 5.

The theorem just proved has begun our study of the consequences of requiring that $(\mathfrak{R}, F)$ take on some of the algebraic characteristics of a field. The rest of the theorems in this section will be concerned with the implications of making $(\mathfrak{R}, F)$ behave more and more like a field. For our next result, we need to recall the *distributive laws*:

$$a \cdot (b + c) = (a \cdot b) + (a \cdot c) \tag{26}$$

for all a, b, c in $\mathfrak{R}$ (the left distributive law), and

$$(a + b) \cdot c = (a \cdot c) + (b \cdot c) \tag{27}$$

for all a, b, c in $\mathfrak{R}$ (the right distributive law).

THEOREM 7. The four conditions (i) $(\mathfrak{R}, F)$ is linear, (ii) $(\mathfrak{R}, +)$ is a group, (iii) $a + b = b + a$ for all a, b in $\mathfrak{R}$, and (iv) $\mathfrak{R}$ satisfies the right distributive law, all hold if and only if Π is $(\mathfrak{L}_S, \mathfrak{L}_S)$-Desarguesian.

Proof. Recall that the statement that a plane is $(\mathfrak{L}, \mathfrak{L})$-Desarguesian means that it is $(P, \mathfrak{L})$-Desarguesian for every $P \mathrm{I} \mathfrak{L}$. Assume first that Π satisfies (i), (ii), and (iii). Then by Theorem 6, we know that Π must be $(Y_S, \mathfrak{L}_S)$-Desarguesian. Now let us show that Π is also $[(0), \mathfrak{L}_S]$-Desarguesian, if Π satisfies (iv), and Theorem 3 will complete our proof. But consider the following mapping g_a, for any a in $\mathfrak{R}$:

$$\begin{array}{ll} (x, y) \rightarrow (a + x, y) & [m, k] \rightarrow [m, -(a \cdot m) + k] \\ (m) \rightarrow (m) & [k] \rightarrow [a + k] \\ Y_S \rightarrow Y_S & \mathfrak{L}_S \rightarrow \mathfrak{L}_S. \end{array} \tag{28}$$

First, let us show that g_a is a collineation for any a of $\mathfrak{R}$. Again the different cases must be analyzed, but the most important is that where P is of the form (x, y) and $\mathfrak{M} = [m, k]$. Then $(x, y) \mathrm{I} [m, k]$ if and only if $y = (x \cdot m) + k$. But $g_a(x, y) \mathrm{I} g_a[m, k] \Leftrightarrow (a + x, y) \mathrm{I} [m, -(a \cdot m) + k]$ and also if and only if

$$y = (a + x) \cdot m - (a \cdot m) + k = a \cdot m + x \cdot m - (a \cdot m) + k = x \cdot m + k.$$

Thus $(x, y) \mathrm{I} [m, k]$ if and only if $g_a(x, y) \mathrm{I} g_a[m, k]$. The other cases are simple, and it is easily seen that g_a is a $((0), \mathfrak{L}_S)$-central collineation satisfying the condition

$$g_a(0, 0) = (a, 0). \tag{29}$$

As in the previous theorem, this implies that Π is $((0), \mathfrak{L}_S)$-transitive. This completes the proof of half of Theorem 7. For the other half of the theorem, assume that Π is $(\mathfrak{L}_S, \mathfrak{L}_S)$-Desarguesian. In particular, then, we can apply the results of Theorem 6, since Π must be $(Y_S, \mathfrak{L}_S)$-Desarguesian. Also, Π must be $((0), \mathfrak{L}_S)$-transitive, and so for any a of $\mathfrak{R}$ there must be a $((0), \mathfrak{L}_S)$-central collineation, g_a, such that $g_a(0, 0) = (a, 0)$. As in the proof of the previous theorem, we can write

$$g_a(x, y) = \big(H_a(x, y), K_a(x, y)\big), \tag{30}$$

where H_a and K_a are functions on $\mathfrak{R}$, $\mathfrak{R}$ to $\mathfrak{R}$, and we wish to explore in more detail the characteristics of H_a and K_a. Since (0) is the center of g_a, we know that $K_a(x, y) = y$ (Why?). Also, as in the proof of Theorem 6, the fact that g_a is a $((0), \mathfrak{L}_S)$-central collineation implies that $H_a(x, y) = H_a(a, 0)$, for all x, y in $\mathfrak{R}$. In particular,

$$g_a(0, x) = (a, x) \tag{31}$$

for any x of $\mathfrak{R}$. Now, for the lines, we have

$$g_a[m, k] = [m, n], \tag{32}$$

for some n in R. Since g_a is a collineation, we can write

$$(0, x)\mathrm{I}[m, k] \Leftrightarrow (a, x)\mathrm{I}[m, n] \Leftrightarrow x = (a \cdot m) + n. \tag{33}$$

But $(0, k)\mathrm{I}[m, k]$, so we can conclude that $k = (a \cdot m) + n$. Since $(R, +)$ is a group by Theorem 6, we can solve this last equation for n and we have $n = -(a \cdot m) + k$, or

$$g_a[m, k] = [m, -(a \cdot m) + k]. \tag{34}$$

Next, let us consider an arbitrary point (x, y) incident with $[m, k]$. Then we have the chain of statements

$$\begin{aligned} y = (x \cdot m) + k &\Leftrightarrow g_a(x, y)\mathrm{I}g_a[m, k] \\ &\Leftrightarrow (H_a(x, 0), y)\mathrm{I}[m, -(a \cdot m) + k] \\ &\Leftrightarrow y = H_a(x, 0) \cdot m - (a \cdot m) + k. \end{aligned}$$

Replacing y by $(x \cdot m) + k$, and canceling k, this last equation becomes

$$x \cdot m = H_a(x, 0) \cdot m - (a \cdot m), \tag{35}$$

for all x, m, a in $\mathfrak{R}$. In (35), let $m = 1$, and we have $x = H_a(x, 0) - a$, or

$$H_a(x, 0) = x + a, \tag{36}$$

for all x in $\mathfrak{R}$.

Using (36) in (35), we have

$$(x \cdot m) + (a \cdot m) = (x + a) \cdot m, \tag{37}$$

for all x, m, a in $\mathfrak{R}$, proving that $(\mathfrak{R}, F)$ satisfies condition (iv). To show that condition (iii) is satisfied and complete the proof of the theorem, we require the following *algebraic* definition and lemmas.

DEFINITION. A linear planar ternary ring in which $(\mathfrak{R}, +)$ is a group and the right distributive law is satisfied is called a right Veblen-Wedderburn (V-W) system.

LEMMA 5. In a right *V-W* system $(-x) \cdot y = -(x \cdot y)$ for all x, y.

Proof. We have the equalities $0 = \big(x + (-x)\big) \cdot y = (x \cdot y) + (-x) \cdot y$, so $-(x \cdot y) = (-x) \cdot y$.

LEMMA 6. If $(\mathfrak{R}, F)$ is a right V-W system, then for any $a \neq 1$, and any b of $\mathfrak{R}$, the equation $(x \cdot a) + b = x$ has a unique solution for x in $\mathfrak{R}$.

Proof. Using condition (c) of Theorem 4.1, and the linearity of $(\mathfrak{R}, F)$, we see that $(x \cdot a) + b = (x \cdot 1) + 0 = x$ has a unique solution x in $\mathfrak{R}$.

LEMMA 7. A right V-W system $\mathfrak{R}$ satisfies $a + b = b + a$ for any a, b of $\mathfrak{R}$.

Proof. For any pair a, b in $\mathfrak{R}$,

$$b + a - b = a \cdot k \tag{38}$$

for some k in $\mathfrak{R}$. Assume $k \neq 1$ for some pair a, b in $\mathfrak{R}$, since otherwise the lemma would be true automatically. Then for that a, b, k, there is a unique solution to $(x \cdot k) + b = x$, by the previous lemma. Thus,

$$-(x \cdot k) + x = b \tag{39}$$

has a unique solution for x. But, using Lemma 5, we can write

$$\begin{aligned} -[(x + a) \cdot k] + (x + a) &= [-(x + a)] \cdot k + x + a \\ &= (-a - x) \cdot k + x + a \\ &= -(a \cdot k) - (x \cdot k) + x + a \\ &= -(a \cdot k) + b + a \\ &= b. \end{aligned}$$

Thus, $x + a$ is also a solution to (39), but since x was a unique solution, we can conclude that $x = x + a$, or $a = 0$. Thus, if $b + a - b = k \cdot a$, $k \neq 1$, then $a = 0$, or, if $a \neq 0$, $b + a = a + b$, proving the lemma and the theorem.

Exercise

10. Complete the case analysis proving that the mapping g_a defined in (28) is a collineation.

In the proof of this last theorem, we have seen even more strongly the connections between the algebraic and geometric notions, especially in the fact that the last condition to be proved in Theorem 7 had to be attacked from a purely algebraic point of view, even though its validity is implied by a geometric condition. In order to complete the picture we have begun, we have to study two more algebraic situations: (1) the left distributive law, and (2) associativity of $(\mathfrak{R}, \cdot)$. The next two theorems complete our discussion.

THEOREM 8. $(\mathfrak{R}, F)$ is linear and $(\mathfrak{R} - \{0\}, \cdot)$ is a group if and only if Π is $((0), [0])$-Desarguesian.

Proof. Assume that $(\mathfrak{R}, F)$ is linear and multiplication is associative. For any a in $\mathfrak{R} - \{0\}$, define the mapping h_a by

$$\begin{array}{ll} (x, y) \to (x \cdot a, y) & [m, k] \to [a^{-1} \cdot m, k] \\ (m) \to (a^{-1} \cdot m) & [k] \to [k \cdot a] \\ Y_S \to Y_S & \mathfrak{L}_S \to \mathfrak{L}_S \end{array} \tag{40}$$

Checking that h_a is a collineation when $(\mathfrak{R}, F)$ is linear and $(\mathfrak{R} - \{0\}, \cdot)$ a group is left as an exercise. The collineation h_a is a $((0), [0])$-central collineation with

$$h_a(1, 0) = (a, 0), \tag{41}$$

and it is easy to deduce that Π must be $((0), [0])$-transitive and thus $((0), [0])$-Desarguesian. Next, assume that Π is $((0), [0])$-transitive. Then for any a in $\mathfrak{R} - \{0\}$, there is a $((0), [0])$-central collineation, g_a, satisfying (41). We can define two one-one transformations S_a and T_a of $\mathfrak{R}$ onto $\mathfrak{R}$ by

$$h_a(m) = (mS_a), \qquad h_a[k] = [kT_a]. \tag{42}$$

In general we shall then have

$$h_a(x, y) = (xT_a, y), \qquad h_a[m, k] = [mS_a, k], \tag{43}$$

since h_a is a $((0), [0])$-central collineation. Note that $1T_a = a$. Now

$$F(x, m, k) = y \Leftrightarrow F(xT_a, mS_a, k) = y, \tag{44}$$

that is, $F(x, m, k) = F(xT_a, mS_a, k)$. Let $k = 0$ to obtain

$$x \cdot m = (xT_a) \cdot (mS_a) \tag{45}$$

for any x, m in $\mathfrak{R} - \{0\}$. Take $x = 1$ in (45) to see that

$$m = a \cdot (mS_a), \tag{46}$$

for any a, m in $\mathfrak{R} - \{0\}$. From (46) for $m = a$, we obtain $a = a \cdot aS_a$ and so $aS_a = 1$. Now, in (45), let $m = a$ to see that $x \cdot a = (xT_a) \cdot (aS_a)$, that is,

$$xT_a = x \cdot a \tag{47}$$

for any x and a in $\mathfrak{R}$. Relation (45) then becomes

$$x \cdot m = (x \cdot a) \cdot (mS_a). \tag{48}$$

Next, in (48), write $m = a \cdot y$, and we have

$$x \cdot (a \cdot y) = (x \cdot a) \cdot ((a \cdot y)S_a) = (x \cdot a) \cdot y, \tag{49}$$

since (46) for $m = a \cdot y$ becomes $a \cdot y = a \cdot [(a \cdot y)S_a]$ and so $(a \cdot y)S_a = y$.

But (49) implies the associative law of multiplication, since it holds for all x, a, y in $\mathfrak{R} - \{0\}$. Since $(\mathfrak{R} - \{0\}, \cdot)$ is a group, (46) implies that $mS_a = a^{-1} \cdot m$. To prove the linearity of $(\mathfrak{R}, F)$, we can rewrite (44) now as $F(x, m, k) = F(x \cdot a, a^{-1} \cdot m, k)$. Let $m = a$, and linearity follows from $F(x, l, k) = x + k$, so that the proof of our theorem is complete.

Exercises

11. Prove that the mapping h_a defined by (40) is a collineation.

12. Show that the ((0), [0])-central collineation h_a satisfying (41) must also satisfy (43).

We come now to the final theorem in this series, and when we have proved it we shall have completed the most difficult part of our present undertaking.

THEOREM 9. The properties (i) $(\mathfrak{R}, F)$ is linear, (ii) $(\mathfrak{R}, +)$ is a group, (iii) $a + b = b + a$ for all a, b in $\mathfrak{R}$, and (iv) $\mathfrak{R}$ satisfies the left distributive law, all hold if and only if Π is (Y_S, Y_S)-Desarguesian.

Proof. It is instructive to note the manner in which this theorem is "dual" to Theorem 7, for here we have replaced right by left distributivity, and $(\mathfrak{L}_S, \mathfrak{L}_S)$-transitivity by (Y_S, Y_S)-transitivity. We shall indicate two proofs of the theorem, the second making use of the principle of duality and the duality observed above. The first proof is standard and similar to several proofs already described in detail, so most of it will be left as an exercise. Consider the mapping φ_a given by

$$\begin{aligned} (x, y) &\to (x, x \cdot a + y) \qquad & [m, k] &\to [a + m, k] \\ (m) &\to (a + m) & [k] &\to [k] \\ Y_S &\to Y_S & \mathfrak{L}_S &\to \mathfrak{L}_S. \end{aligned} \tag{50}$$

Then φ_a is a $(Y_S, [0])$-central collineation with $\varphi_a(0) = (a)$, and $(Y_S, [0])$-transitivity follows, and finally also (Y_S, Y_S)-transitivity by the dual of Theorem 3. The other half of the theorem is proved by letting φ_a be the $(Y_S, [0])$-central collineation with $\varphi_a(0) = (a)$, and a proof is obtained with some effort along standard lines.

The second proof is much more interesting in that it is different from what we have been doing throughout this section, and also because it makes heavy use of our study of duality. The idea of this second proof is to properly coordinatize the dual plane, Π^*, of Π and then make use of what we have already proved. In particular, we would like to coordinatize Π^* so that the line $[m, k]$ of Π becomes the point $(m, k)'$ of Π^*, the line $[k]$ of Π becomes the point $(-k)'$ of Π^*, and $\mathfrak{L}_S$ in Π becomes the point Y_s' in Π^*.

Similarly, the points (x, y), (m), and Y_S of Π become, respectively, the lines $[-x, y]'$, $[m]'$, and $\mathfrak{L}'_S$ of Π^*. We leave to the reader the determination of the coordinatizing quadrangle of Π^* which will allow us to name the elements of Π^* as indicated. The ternary function, G, of Π^* is now determined by the property

$$(x, y)' \mathrm{I} [-m, k]' \Leftrightarrow y = G(x, -m, k). \tag{51}$$

But $(x, y)' \mathrm{I} [-m, k]' \Leftrightarrow (m, k) \mathrm{I} [x, y] \Leftrightarrow k = F(m, x, y)$. Thus, we have the relation

$$y = G(x, -m, k) \Leftrightarrow k = F(m, x, y) \tag{52}$$

for all x, m, k, y in $\mathfrak{R}$. We can now complete the proof of the theorem. If we assume that Π is (Y_S, Y_S)-Desarguesian, then, using Theorem 6, we know that $(\mathfrak{R}, F)$ is linear and $(\mathfrak{R}, +)$ is a group. But, using Exercise 6, we know that Π^* is $(\mathfrak{L}'_S, \mathfrak{L}'_S)$-Desarguesian, so we can apply the results of Theorem 7 to $(\mathfrak{R}, G)$. To do this, let $\oplus$ and $\circ$ be the operations in $(\mathfrak{R}, G)$ defined by

$$x \oplus y = G(x, 1, y) \tag{53}$$

$$x \circ y = G(x, y, 0).$$

Thus, $(\mathfrak{R}, \oplus)$ is a group, and $(\mathfrak{R}, G)$ satisfies the right distributive law. Since $(\mathfrak{R}, F)$ and $(\mathfrak{R}, G)$ are both linear, (52) can be rewritten as:

$$y = [x \circ (-m)] \oplus k \Leftrightarrow k = (m \cdot x) + y, \tag{54}$$

and thus

$$[x \circ (-m)] \oplus k = -(m \cdot x) + k. \tag{55}$$

Let $k = 0$, and we have

$$x \circ (-m) = -(m \cdot x). \tag{56}$$

Next, let $m = -1$ in (55), and obtain $x \oplus k = x + k$. Thus, $(\mathfrak{R}, \oplus)$ and $(\mathfrak{R}, +)$ are isomorphic groups, and we can write (55) as $x \circ (-m) = (-m) \cdot x$ for all x, m in $\mathfrak{R}$ and thus

$$x \circ m = m \cdot x \tag{57}$$

for all x, m in $\mathfrak{R}$.

If we now use the result of Theorem 7 that $(\mathfrak{R}, G)$ satisfies the right distributive law and substitute this result in (57), we have $x \circ (a + b) = (a + b) \cdot x$, while

$$x \circ (a + b) = (x \circ a) + (x \circ b) = (a \cdot x) + (b \cdot x).$$

Thus, $(a + b) \cdot x = (a \cdot x) + (b \cdot x)$, which is the desired distributive law in $(\mathfrak{R}, F)$. Reversing the argument is similar: Assuming $(\mathfrak{R}, F)$ satisfies the conditions of the theorem and having coordinatized Π^* properly, we can show that $(\mathfrak{R}, G)$ must satisfy the conditions of Theorem 7, so Π^* is $(\mathfrak{L}'_S, \mathfrak{L}'_S)$-Desarguesian, which implies that Π is (Y_S, Y_S)-Desarguesian.

Exercises

13. Complete the first proof of Theorem 9.

14. In the second proof of the theorem, what is the proper coordinatizing quadrangle for Π^* that will yield the stated coordinates for Π^*?

15. Complete the second half of the second proof of Theorem 9.

We have come to the end of the so-called coordinatization theorems. It is now an easy matter, which we defer to the next chapter, to put all these theorems together and prove the "Fundamental Theorem" of projective geometry. The essence of the matter—the relationship between algebra and geometry—is contained in this chapter. It is interesting for its own sake to see the geometric structure of the plane grow as the algebraic structure of one of its coordinate systems is made more field-like. One natural question arising from the theorems in this chapter is to characterize and classify those algebraic systems that satisfy the conditions of the theorems, and to give specific examples of the algebraic systems and study them. Much work has been done along these lines and much remains to be done.

[6]

The Fundamental Theorem

With the difficult theorems behind us we shall turn, in this chapter, to tying together the results of the previous chapters into a proof of the "Fundamental Theorem of Finite Projective Geometry." This theorem asserts that *a finite projective plane is Desarguesian if and only if it is a field plane.* To prove this theorem, we need first to examine the field planes more closely to determine the exact nature of their coordinatizing planar ternary rings.

1. Coordinates in a Field Plane

In Section 3.5 we began to investigate another way of viewing the triples of field elements which we used to describe the field planes. We would like to carry this analysis further and prove that if $(\mathfrak{R}, F)$ is *any* coordinatizing *ternary* ring of a field plane $\Sigma = PG(2, p^n)$, then $(\mathfrak{R}, F)$ is linear, and the *binary* ring $(\mathfrak{R}, +, \cdot)$ is isomorphic with the original field used to define Σ. The easiest way of proving this is to show that the plane $PG(2, p^n)$ is isomorphic to the plane Π whose coordinatizing ternary ring is linear and is constructed out of the finite field $GF(p^n)$ as in Exercise 4.8. (Of course, Theorem 4.2 and Exercise 4.8 completely describe Π in terms of $GF(p^n)$.) We state the result as the first theorem of this chapter.

THEOREM 1. The plane $PG(2, p^n)$ is isomorphic with the plane Π whose coordinate system is linear and is isomorphic with $GF(p^n)$ when considered as a binary ring.

Proof. Consider the mapping of points and lines defined by

$$\begin{aligned} f(x, y, e) &= (x, y) \\ f(e, m, 0) &= (m) \\ f(0, e, 0) &= Y_S \end{aligned} \tag{1}$$

for all x, y, m in $GF(p^n)$, and

$$f\begin{pmatrix} m \\ -e \\ k \end{pmatrix} = [m, k]$$

$$f\begin{pmatrix} -e \\ 0 \\ k \end{pmatrix} = [k] \qquad (2)$$

$$f\begin{pmatrix} 0 \\ 0 \\ e \end{pmatrix} = \mathfrak{L}_S,$$

for all m, k in $GF(p^n)$.

Clearly, f is one-one. To show that f is an isomorphism, we have to show that $P \mathrm{I} \mathfrak{L}$ if and only if $f(P) \mathrm{I} f(\mathfrak{L})$. If $P = (x, y, e)$ and

$$\mathfrak{L} = \begin{pmatrix} m \\ -e \\ k \end{pmatrix},$$

then we have the equivalences $P \mathrm{I} \mathfrak{L} \Leftrightarrow (x \cdot m) - y = k \Leftrightarrow y = (x \cdot m) + k \Leftrightarrow (x, y) \mathrm{I} [m, k]$. If $P = (x, y, e)$ and

$$\mathfrak{L} = \begin{pmatrix} -e \\ 0 \\ k \end{pmatrix},$$

then $P \mathrm{I} \mathfrak{L} \Leftrightarrow -x + k = 0 \Leftrightarrow x = k \Leftrightarrow (x, y) \mathrm{I} [k]$. The remaining cases are similarly checked and f is easily seen to be an isomorphism, as asserted.

Exercise

1. Complete the case analysis showing that f is an isomorphism.

As a corollary to Theorems 3.3, 4.5, and 6.1, we can state the following theorem which, because of Theorem 4.5, asserts the existence of many collineations in planes coordinatized by the most "structured" algebraic systems.

THEOREM 2. Let Π be a plane coordinatized by a linear planar ternary ring whose associated binary ring is the finite field $GF(p^n)$. Then the binary ring defined by any coordinatizing ternary ring of Π is isomorphic with $GF(p^n)$.

The theorem implies that, given any two ordered quadrangles in such a plane, say $\{X_1, X_2, X_3, X_4\}$ and $\{Y_1, Y_2, Y_3, Y_4\}$, there is a collineation f mapping $X_i \rightarrow Y_i$ for $i = 1, 2, 3, 4$. It is of some interest to turn the question around and ask: If Π is a finite plane with this property (that given any two ordered quadrangles there is a collineation mapping one into the other) is there any conclusion we can reach about the planar ternary rings coordinatizing Π? This is related to the questions we asked (and answered) in the previous chapter, but the collineations assumed there were all central collineations, while here we make no such assumptions. Nevertheless, there is a good deal that can be said, and even more than we asked for. One of the more important theorems is due to T. G. Ostrom and A. Wagner, and we state it now without proof.

THEOREM 3. Let Π be a finite projective plane with the property that, given any two ordered pairs of points $\{P_1, P_2\}$ and $\{Q_1, Q_2\}$, there is a collineation f of Π such that $f(P_1) = Q_1$ and $f(P_2) = Q_2$. Then Π is a field plane.

Notice how this theorem more than answers the question we asked, and even gives a new way of characterizing finite field planes as finite planes whose collineation groups are "transitive" on ordered pairs of points.

2. Wedderburn's Theorem

In order to prove the fundamental theorem, we need one more algebraic theorem, which we shall *discuss* in this section. An examination of the theorems of Chapter V will show that we have dealt with every one of the conditions defining a field except that of the commutativity of multiplication. This is due to the fact that *there exist infinite algebraic systems satisfying all of the conditions except commutativity of multiplication*, and in the statement of the fundamental theorem for infinite planes this must be taken into account. The situation in the finite case, however, is made much neater by the following celebrated theorem due to Wedderburn:

THEOREM 4. Let $\mathfrak{F}$ be a finite set with two binary operations $+$ and $\cdot$ defined on it such that conditions (a), (b), (c), (e), and (f) of section 3.1 are all satisfied. Then condition (d) must also be satisfied, that is, $\mathfrak{F}$ must be a finite field.

This theorem states that if a finite algebraic system satisfies all the requirements for being a field except perhaps commutativity of multiplication, then it must be a finite field. There are some similarities between Wedderburn's theorem and Lemma 5.7, but Wedderburn's theorem is much more difficult to prove, and is discussed in most texts on algebra* so we shall omit its proof.

3. The Fundamental Theorem

We have already provided the difficult basic background for our fundamental theorem; in this section we shall simply quote the theorems that provide its proof.

THEOREM 5. Let Π be a finite projective plane. Then Π is Desarguesian if and only if it is a field plane (that is, if its ternary ring is linear and has a finite field as its associated binary ring).

Proof. Assume first that Π is Desarguesian, that is, Π is $(P, \mathfrak{L})$-Desarguesian for every point P and line $\mathfrak{L}$. Choose any coordinatizing quadrangle in Π $\{X, Y, Q, I\}$, and let $(\mathfrak{R}, F)$ be the ternary ring defined. Theorems 5.6, 5.7, 5.8, and 5.9 imply that $(\mathfrak{R}, F)$ is linear and satisfies all the necessary conditions so that the binary ring $(\mathfrak{R}, +, \cdot)$ is a finite field except perhaps for the commutativity of multiplication. But Wedderburn's theorem now provides the implication that $(\mathfrak{R}, +, \cdot)$ is a finite field. Next, assume that Π has as one of its coordinate systems a linear ternary ring whose associated binary ring is finite field $GF(p^n)$. Then Theorem 2 implies that *every* coordinatizing ternary ring $(\mathfrak{R}, F)$ yields a finite field $(\mathfrak{R}, +, \cdot)$. Given any point P and line $\mathfrak{L}$, if $P \mathrel{I} \mathfrak{L}$, choose a coordinatizing quadrangle such that $P = Y_S$, $\mathfrak{L} = \mathfrak{L}_S$. If $P \not\mathrel{I} \mathfrak{L}$, choose a quadrangle such that $P = (0)$ and $\mathfrak{L} = [0]$. Then Theorems 5.6 and 5.8 imply $(P, \mathfrak{L})$-transitivity for every point P and line $\mathfrak{L}$.

Since every finite field has prime-power order, a consequence of the fundamental theorem is that every Desarguesian plane has prime-power order. Much more is true, however, as it is possible to prove that every V-W system has prime-power order. This is not too difficult to prove, but it requires a little more group theory than we have discussed so far. These results are all partial solutions of the general problem of the possible orders of finite planes discussed in Chapter 2.

* (See for example, I. N. Herstein, *Topics in Algebra*, pp. 314–324.)

4. Pappus' Property

In the fundamental theorem we have obtained an algebraic characterization of one of the most important geometric properties that can exist in a projective plane. There are many such conditions that can be studied, and in this section we shall study another well-known geometric property. Our interest in Pappus' property is both in its relation to Desargues' property and in the fact that it gives us another striking example of the beautiful interactions between algebra and geometry.

The geometric property we shall study now is the following: Let Π be a projective plane, and let $\mathfrak{L}_1$, $\mathfrak{L}_2$ be any two distinct lines of Π with $\mathfrak{L}_1 \cap \mathfrak{L}_2 = P$. Assume also that P_1, P_2, P_3 are three points incident with $\mathfrak{L}_1$, and that Q_1, Q_2, Q_3, are three points incident with $\mathfrak{L}_2$, such that none of the P_i or Q_j is the point P. Then Π is said to satisfy the Pappus property if for any such choice of $\mathfrak{L}_1, \mathfrak{L}_2, P_i, Q_j$ the points

$$\begin{aligned} R_1 &= [P_1 \cdot Q_2] \cap [Q_1 \cdot P_2], \\ R_2 &= [P_1 \cdot Q_3] \cap [Q_1 \cdot P_3], \\ R_3 &= [P_2 \cdot Q_3] \cap [Q_2 \cdot P_3], \end{aligned} \tag{3}$$

are collinear.

THEOREM 6. Any plane which has the Pappus property is Desarguesian.

Proof. Let $\{A_1, A_2, A_3\}$ and $\{B_1, B_2, B_3\}$ be any two triangles perspective from a point P. Then we wish to show that they are perspective from some line $\mathfrak{L}$. To restate the problem, let $\mathfrak{L}$ be determined by the points $[A_1 \cdot A_2] \cap [B_1 \cdot B_2]$ and $[A_1 \cdot A_3] \cap [B_1 \cdot B_3]$, and we must prove that $\mathfrak{L}$ contains the point $[A_2 \cdot A_3] \cap [B_2 \cdot B_3]$. To begin with, let

$$\mathfrak{M} = [A_1 \cdot (\mathfrak{L} \cap [A_2 \cdot B_2])] \tag{4}$$

and

$$P_1 = \mathfrak{M} \cap [B_1 \cdot B_3] \tag{5}$$

(see Fig. 6.1).

Also, let

$$\begin{aligned} P_2 &= [P_1 \cdot B_2] \cap [A_1 \cdot A_2], \\ P_3 &= [A_1 \cdot P_1] \cap [A_3 \cdot B_3]. \end{aligned} \tag{6}$$

We see that the triples P, B_1, A_1 and P_1, P_2, B_2 are collinear on the lines $[A_1 \cdot B_1]$ and $[P_1 \cdot P_2]$, respectively (Why?). Applying Pappus' property, we

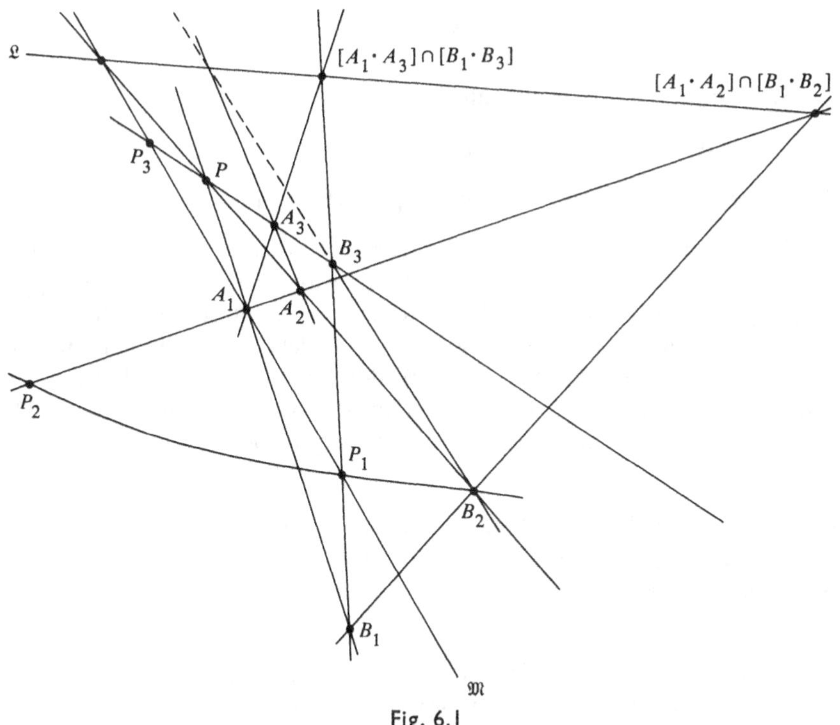

Fig. 6.1

see that the points $([P \cdot P_2] \cap [P_1 \cdot B_1])$, $([P \cdot B_2] \cap [P_1 \cdot A_1])$, and $([B_1 \cdot B_2] \cap [P_2 \cdot A_1])$ are collinear, and, since the last two points are on $\mathfrak{L}$, we have

$$([P \cdot P_2] \cap [P_1 \cdot B_1]) \mathrm{I} \mathfrak{L} \tag{7}$$

Next, consider the two triples P_2, A_2, A_1 and A_3, P_3, P. Using the Pappus property and (7), we can deduce that

$$([P_2 \cdot P_3] \cap [A_2 \cdot A_3]) \mathrm{I} \mathfrak{L} \tag{8}$$

(Why?). Finally, using the two triples P, B_3, P_3 and P_1, P_2, B_2, we deduce that

$$([P_2 \cdot P_3] \cap [B_2 \cdot B_3]) \mathrm{I} \mathfrak{L}. \tag{9}$$

This almost completes our proof of the theorem. Have you noticed why this argument is not valid if $P \mathrm{I} \mathfrak{L}$? We have actually proved that Π is $(P, \mathfrak{L})$-Desarguesian for all $P \not{\mathrm{I}} \mathfrak{L}$. To complete the proof, we need the following lemma whose proof is similar to that of Theorem 5.3.

LEMMA 1. Let Π be $(P_1, \mathfrak{L})$ and $(P_2, \mathfrak{L})$-transitive for $P_i \not{\mathrm{I}} \mathfrak{L}$. Then Π is $(Q, \mathfrak{L})$-transitive, where $Q = ([P_1, P_2] \cap \mathfrak{L})$.

Exercises

2. Why are (7), (8), and (9) valid?
3. Where does the proof fail if $P \mathrm{I} \mathfrak{L}$?
4. Prove Lemma 1, and use it to complete the proof of the theorem.

Next, we shall consider the algebraic consequences of the Pappus property. Of course, Theorem 6 implies that any plane having the Pappus property is coordinatized by a linear ternary ring, whose associated binary ring satisfies all the defining conditions of a field except perhaps that of commutative multiplication. Also, Wedderburn's theorem allows us to assert that *any finite plane having the Pappus property is a field plane.* The next theorem clarifies the situation for arbitrary planes.

THEOREM 7. Let Π be a Desarguesian projective plane. Then Π satisfies Pappus' property if and only if multiplication in the coordinatizing ring of Π is commutative.

Proof. The proof of this theorem will, strictly speaking, be valid only for finite planes, but it is also true for infinite planes since all of the theorems quoted in the proof hold for infinite as well as finite planes. In particular, all the results of Section 3.4 concerning the collineations of finite field planes can be carried over for planes defined over infinite systems satisfying all the conditions of fields (except perhaps that of commutative multiplication). Also, the theorems of Chapter 5 were all proved without reference to the finiteness of the planes studied there. The theorem is proved by choosing the *proper* coordinatizing quadrangle in the plane Π. We are, of course, using the fact that all the coordinate systems in a Desarguesian plane are isomorphic. Given any two triples satisfying the hypotheses of Pappus' condition, P_1, P_2, P_3, and Q_1, Q_2, Q_3, let (see Fig. 6.2)

$$\begin{aligned}(0, 0) &= [P_1 \cdot P_2] \cap [Q_1 \cdot Q_2]\\ (0) &= P_1\\ Y_S &= Q_2\\ (1, 1) &= [P_1 \cdot Q_1] \cap [P_3 \cdot Q_2].\end{aligned} \tag{10}$$

Since no three of these points can be collinear (Why?), we can use this quadrangle to coordinatize Π. We can deduce easily that

$$(1, 0) = P_3, \qquad (0, 1) = Q_1. \tag{11}$$

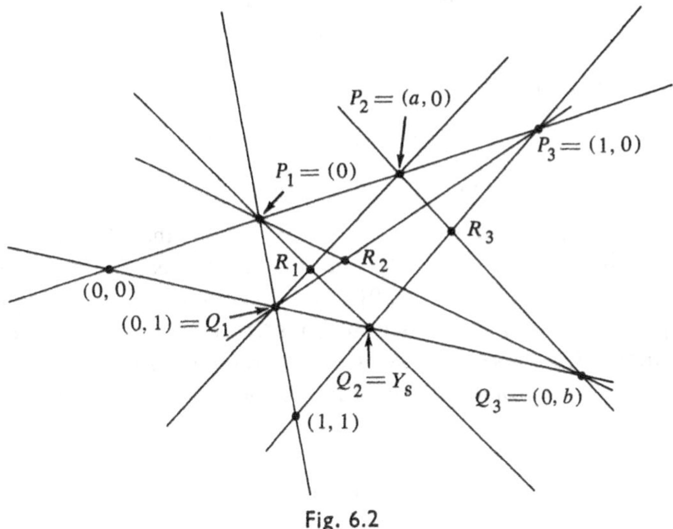

Fig. 6.2

It follows that, for *some elements* a, b ($\neq 0, 1$) in the coordinate system, we have

$$(a, 0) = P_2, \qquad (0, b) = Q_3. \tag{12}$$

Consider now the three points R_1, R_2, R_3 defined in (3) above. It is not difficult to show that

$$\begin{aligned} R_1 &= (-a^{-1}), \\ R_2 &= (1 - b, b), \\ R_3 &= (1, b - a^{-1}b). \end{aligned} \tag{13}$$

For example, to show that R_3 is as asserted,

$$R_3 = [Y_S \cdot (1, 1)] \cap [(a, 0) \cdot (0, b)] = (1, \alpha)$$

for some α, since $[Y_S \cdot (1, 1)] = [1]$. But $[(a, 0) \cdot (0, b)] = [m, k]$, so $0 = (a \cdot m) + k$ and $b = k$, hence we conclude that $[m, k] = [-a^{-1} \cdot b, b]$. Using this to solve for α, we see that $(1, \alpha) \mathrm{I} [m, k]$ implies that

$$\alpha = (1 \cdot m) + k = (-a^{-1} \cdot b) + b.$$

Similarly, it can be shown that $R_1 = (-a^{-1})$ and $R_2 = (1 - b, b)$. Now, to determine when R_1, R_2, and R_3 are collinear, we notice that

$$\begin{aligned} [R_1 \cdot R_2] &= [-a^{-1}, k_1], \\ [R_1 \cdot R_3] &= [-a^{-1}, k_2], \end{aligned} \tag{14}$$

for some k_1, k_2. Restating this, $R_2 \text{I} [-a^{-1}, k_1]$ is equivalent to

$$b = (-(1 - b) \cdot a^{-1}) + k_1,$$

and $R_3 \text{I} [-a^{-1}, k_2]$ is equivalent to $b - (a^{-1} \cdot b) = -a^{-1} + k_2$. Thus,

$$\begin{aligned} k_1 &= b + a^{-1} - (b \cdot a^{-1}), \\ k_2 &= b + a^{-1} - (a^{-1} \cdot b). \end{aligned} \tag{15}$$

Then $k_1 = k_2$ if and only if $b \cdot a^{-1} = a^{-1} \cdot b$ and thus $a \cdot b = b \cdot a$. But $k_1 = k_2$ if and only if R_1, R_2, and R_3 are collinear. Thus, if for any choice of a, b (that is, of P_2 on [0, 0] and Q_3 on [0]) we have $k_1 = k_2$ (that is, if the Pappus property holds), then the coordinatizing ring has commutative multiplication. Conversely, if we have the property that $a \cdot b = b \cdot a$ for any a, b, then $k_1 = k_2$ for all choices of P_2, Q_3 and the Pappus property follows.

Exercise

5. Verify the forms of R_1 and R_2 as given in (13).

Theorem 7 can be used to show that there exist projective planes that are Desarguesian but do not satisfy the Pappus property, since algebraic systems can be constructed that satisfy all of the conditions of a field *except* commutativity of multiplication. Wedderburn's theorem, however, implies that such a plane must be infinite, and as a corollary to Wedderburn's theorem and Theorem 7 we have the following remarkable result.

THEOREM 8. Any finite Desarguesian plane satisfies the Pappus property.

This theorem is remarkable in that the purely algebraic Wedderburn theorem is a key step in its proof, while no mathematician has yet been able to devise a geometrical proof (along the lines of the proof of Theorem 6, for example). Such a proof would be of great interest, but no one has seen a way of making use of the finiteness of Π geometrically, even though a finiteness condition is quite well understood for algebraic systems and can be used to prove very strong results, such as the Wedderburn theorem.

[7]

Some Non-Desarguesian Planes

In the previous chapters, we have proved several theorems about projective planes and we have also constructed an infinitely large class of finite planes. One disturbing fact, however, is that the only planes we have *explicitly* constructed (either directly or by means of their coordinate rings) have all been Desarguesian, while the interesting theorems of Chapter 5 were concerned with planes that are not necessarily Desarguesian. A natural question that should have occurred to the reader by now concerns the existence and variety of non-Desarguesian planes. The answer is that there are a great many non-Desarguesian planes of various "types" as characterized by the hypotheses of the theorems of Chapter 5. In this chapter, we shall first examine more closely the finite fields and then use our results to construct some examples of non-Desarguesian planes.

1. Subfields and Automorphisms of Finite Fields

This section will explore some of the simple facts about finite fields we shall require for the remainder of the chapter. First, consider $\mathfrak{F} = GF(p^a)$, the finite field containing p^a elements. In $\mathfrak{F}$ there is a multiplicative identity, e. Now define

$$e_1 = e, \qquad e_n = e_{n-1} + e \qquad \text{for } n > 1. \tag{1}$$

Then we can easily prove the following result

LEMMA 1. For any integers m, $n \geq 1$, $e_m + e_n = e_{m+n}$ and $e_m e_n = e_{mn}$. Also, $e_m - e_n = e_{m-n}$ if $m > n$.

Proof. A simple induction suffices to prove the first assertion. First, $e_m + e_1 = e_{m+1}$. Next, assume $e_m + e_i = e_{m+i}$ for $i < n$. Then, since $e_n = e_{n-1} + e$, $e_m + e_n = e_m + e_{n-1} + e = e_{m+n-1} + e = e_{m+n}$. The last part of the lemma follows from the first (Why?). The middle statement is proved inductively as is the first, and we leave it as an exercise.

Exercise

1. Show that $e_m \cdot e_n = e_{mn}$ for $m, n \geq 1$.

Now consider the sequence $e_1, e_2, \cdots, e_n, \cdots$. Since $\mathfrak{F}$ has only a finite number of elements, the sequence must have repetitions. Thus, for some i, j $(i < j)$, we must have $e_i = e_j$, or

$$e_{j-i} = 0 \tag{2}$$

Thus, for some $n > 0$, $e_n = 0$. Let q_0 be the *least* integer satisfying the condition $e_q = 0$. Then, by Lemma 1, the sequence $e_1, \cdots, e_n, \cdots$, will be of the form $e, e_2, \cdots, e_{q_0-1}, 0, e, e_2, \cdots$, repeating the basic sequence over and over. Our next result is stated as a theorem, but it will be proved in a series of lemmas.

THEOREM 1. The system consisting of the set $0, e, e_2, \ldots, e_{q_0-1}$, and the operations $+$ and $\cdot$ of $\mathfrak{F}$, forms a field isomorphic to the prime field $GF(p)$.

Proof. We shall prove first that q_0 is a prime number, then that the set forms a field $GF(q_0)$, and finally show that $q_0 = p$, and the proof will be complete. To begin, we have Lemma 2.

LEMMA 2. The integer q_0 is prime.

Proof. Assume $q_0 = mn$ for some integers m and n. Then by Lemma 1,

$$e_m \cdot e_n = e_{mn} = e_{q_0} = 0. \tag{3}$$

Since $\mathfrak{F}$ is a field, $e_m \cdot e_n = 0$ implies $e_m = 0$ or $e_n = 0$ (Why?). But $m < q_0$ and $n < q_0$, and q_0 was chosen to be the *least* integer such that $e_q = 0$. Thus, (3) cannot hold for proper divisors of q_0, so q_0 must have no proper divisors, that is, q_0 is a prime.

LEMMA 3. The set $\mathfrak{G} = 0, e, e_2, \ldots, e_{q_0-1}$ forms a subfield of $\mathfrak{F}$.

Proof. Lemma 1 shows directly that $\{\mathfrak{G}, +\}$ is a group, while $\{\mathfrak{G}, +\}$ and $\{\mathfrak{G}, \cdot\}$ must be associative since $\{\mathfrak{F}, +\}$ and $\{\mathfrak{F}, \cdot\}$ are. Also, the distributive laws are demonstrated to hold by using Lemma 1. Showing that every element of $\mathfrak{G} - \{0\}$ has a multiplicative inverse is slightly harder, but can be done in the same way as for the prime fields (which is not surprising, since we are looking at the prime field $GF(q_0)$).

Exercise

2. Fill in the gaps in the proof of Lemma 3 and complete the proof that $\mathfrak{G}$ is isomorphic with $GF(q_0)$.

The final result, that $q_0 = p$ follows from the fact that $\{\mathfrak{G}, +\}$ is a *subgroup* of $\{\mathfrak{F}, +\}$, and the following famous result.

LEMMA 4. Let $\mathfrak{G}_1$, $\mathfrak{G}_2$ be two finite groups with $\mathfrak{G}_1$ a subgroup of $\mathfrak{G}_2$. Then the order of $\mathfrak{G}_1$, m, is a divisor of the order of $\mathfrak{G}_2$, n.

This lemma proves $q_0 = p$, for it asserts that q_0 divides p^a. But p is the only prime number that can divide p^a, so $q_0 = p$ follows.

Proof of the Lemma. For any element x in $\mathfrak{G}_2$, let $x\mathfrak{G}_1 = \{y \mid y = xg_1$ for some g_1 in $\mathfrak{G}_1\}$. Then $x\mathfrak{G}_1$ is called a *coset of* $\mathfrak{G}_1$ *in* $\mathfrak{G}_2$. Since $\mathfrak{G}_2$ is a group, $x\mathfrak{G}_1$ has as many elements in it as does $\mathfrak{G}_1$ (Why?). Also, if two cosets, $x_1\mathfrak{G}_1$ and $x_2\mathfrak{G}_1$ have a single element in common, they are the same coset. For, if $x_1g_1 = x_2g_1'$, then $x_1 = x_2g_1'g_1^{-1} = x_2g_2$, where $g_2 = g_1'g_1^{-1}$ is in $\mathfrak{G}_1$. Now, let g be in $x_1\mathfrak{G}_1$. Then $g = x_1g_3$, for some g_3 in $\mathfrak{G}_1$. But then $g = x_2g_2g_3 = x_2(g_2g_3)$ is in $x_2\mathfrak{G}_1$. Thus $x_1\mathfrak{G}_1 = x_2\mathfrak{G}_1$. Now, since every element x of $\mathfrak{G}_2$ is in some coset, namely $x\mathfrak{G}_1$, of $\mathfrak{G}_1$, there must be k (≥ 1) *disjoint* cosets whose set-theoretic union is all of $\mathfrak{G}_2$. Thus, $mk = n$, and the lemma is proved.

As a corollary to Theorem 1, we have some important information about addition in any finite field. If we define, for any a in $GF(p^n)$,

$$1a = a, \qquad ia = (i - 1)a + a, \tag{4}$$

we see that $ia = a + a + \cdots + a$, where a is added to itself $i - 1$ times. Thus, $ia = (e + e + \cdots + e)a = e_i a$. From this, we can conclude that, since $e_p = 0$,

$$pa = 0, \tag{5}$$

for any a in $GF(p^n)$.

Keeping this last fact in mind, we proceed to the definition of an important mapping in $GF(p^n)$. For any a in $GF(p^n)$, we define the powers a^i of a for $i > 1$ by

$$a^1 = a, \qquad a^i = a^{i-1}a. \tag{6}$$

We shall prove the following theorem, which will enable us to construct *algebraically* some important non-Desarguesian planes:

THEOREM 2. Define a mapping of $GF(p^n)$ onto itself by $x \to x^p$ for every x in $GF(p^n)$. Then this mapping is an automorphism of $GF(p^n)$.

Proof. First, $(xy)^p = x^p y^p$ in any field. The binomial theorem states that $(x + y)^p = x^p + px^{p-1}y + \cdots + pxy^{p-1} + y^p$, where the coefficients are to be interpreted as in (4). Since all of the coefficients, except those of x^p and y^p, have p as a factor, (5) implies that they are all zero, and so $(x + y)^p = x^p + y^p$. Next, if $x^p = y^p$, then $0 = x^p - y^p = (x - y)^p$. But if a power of any element is zero, the element itself must be zero (Why?), and we can conclude that $x = y$, that is, the mapping is one-one. Finally, any one-one mapping of a finite set into itself must be onto, and we have completed the proof.

2. The Algebras $\mathfrak{A}_\delta$

In this section we shall define and discuss an infinite class of finite coordinatizing ternary rings of non-Desarguesian projective planes. Let $\mathfrak{F} = GF(p^n)$, where p and n are arbitrary, except that we stipulate $n \geq 2$. Then consider the set $\mathfrak{A}$ of ordered pairs (x_1, x_2) of elements of $\mathfrak{F}$. Clearly, $\mathfrak{A}$ has $p^{2n} = (p^n)^2$ elements (Why?). We define two operations $\oplus$ and $\circ$ in $\mathfrak{A}$ by

$$(x_1, x_2) \oplus (y_1, y_2) = (x_1 + y_1, x_2 + y_2) \tag{7}$$

$$(x_1, x_2) \circ (y_1, y_2) = (x_1 y_1 + \delta x_2{}^p y_2, x_1{}^p y_2 + x_2 y_1) \tag{8}$$

for some fixed δ in $\mathfrak{F}$, and call $\mathfrak{A}$, *together with these two operations*, $\mathfrak{A}_\delta$. We shall attempt to prove that $\mathfrak{A}_\delta$ is the binary ring defined by a linear planar ternary ring for suitably chosen elements δ. The first fact to observe is that $(\mathfrak{A}_\delta, \oplus)$ forms a commutative group whose identity element is $(0, 0)$ (Why?). To show that $(\mathfrak{A}_\delta - \{(0, 0)\}, \circ)$ is a loop is somewhat more difficult. We begin by noticing that $(e, 0)$ is an identity under $\circ$. Next, let us examine the conditions under which the product of two elements of $\mathfrak{A}_\delta$ is $= (0, 0)$. For if $(\mathfrak{A} - \{(0, 0)\}, \circ)$ is to be a loop, such a product can never equal the additive identity. From (8) if $(x_1, x_2) \circ (y_1, y_2) = (0, 0)$, we have the two equations

$$x_1 y_1 + \delta x_2{}^p y_2 = 0 \tag{9}$$

and

$$x_1{}^p y_2 + x_2 y_1 = 0. \tag{10}$$

Clearly we must stipulate $\delta \neq 0$, for if $\delta = 0$, set $x_1 = y_1 = 0$ and the product is zero for any values of x_2 and y_2. Assuming, then, that $\delta \neq 0$, when can (9) and (10) hold simultaneously for $(x_1, x_2) \neq (0, 0) \neq (y_1, y_2)$? This cannot hold when $x_2 = 0$, for $x_2 = 0$ implies that $x_1 y_1 = 0 = x_1{}^p y_2$ (Why?). Thus, either $x_1 = 0$, in which case $(x_1, x_2) = (0, 0)$, or if $x_1 \neq 0$, $(y_1, y_2) = (0, 0)$ (Why?). Thus, we can assume $x_2 \neq 0$, and solve (10) for y_1:

$$y_1 = -x_2{}^{-1} x_1{}^p y_2. \tag{11}$$

Using this expression in (9), we have

$$\delta x_2{}^p y_2 = x_2{}^{-1} x_1^{p+1} y_2,$$

that is,

$$\delta x_2^{p+1} = x_1^{p+1} \tag{12}$$

We are allowed to cancel y_2, since by an argument similar to the previous argument y_2 cannot be zero. But (12) can be written

$$\delta = (x_2{}^{-1}x_1)^{p+1} = x^{p+1}, \qquad x = x_2{}^{-1}x_1. \tag{13}$$

Thus, if δ can be chosen in $\mathfrak{F}$ so that it is not a $(p+1)$st power of any element of $\mathfrak{F}$, then (13) states that (9) and (10) can never be simultaneously satisfied unless $(x_1, x_2) = (0, 0)$, $(y_1, y_2) = (0, 0)$, or both are $(0, 0)$. It is not a trivial fact to demonstrate, but it follows from the well-known fact that *the multiplicative group of any finite field is a cyclic group* that, if $p = 2$ and $n > 1$, such a δ always exists. (Can you prove this, assuming the fact that $(\mathfrak{F} - \{0\}, \cdot)$ is cyclic?) To complete the proof that $(\mathfrak{A}_\delta - \{(0, 0)\}, \circ)$ is a loop, using the fact that $(x_1, x_2) \circ (y_1, y_2) = (0, 0)$ if and only if $(x_1, x_2) = (0, 0)$ or $(y_1, y_2) = (0, 0)$, requires a simple counting argument and is left as an exercise. We also leave as an exercise the verification that *both distributive laws* are satisfied by $\mathfrak{A}_\delta$. Finally, if $\mathfrak{A}_\delta$ has associative multiplication, then Wedderburn's theorem will imply that multiplication is commutative and $\mathfrak{A}_\delta$ will be the field $GF(p^{2n})$. Thus, if multiplication is not commutative, it cannot be associative either. Now, the two elements $(0, 1)$, $(0, x)$ commute if and only if $x^p = x$, for $(0, 1) \circ (0, x) = (\delta x, 0)$, while $(0, x) \circ (0, 1) = (\delta x^p, 0)$. To prove that not all elements in $\mathfrak{A}_\delta$ commute, it will suffice to prove that

$$x^p - x = 0 \tag{14}$$

cannot be satisfied for all x in $\mathfrak{F}$. The division algorithm and remainder theorem for polynomials with real coefficients actually hold for polynomials with coefficients in an arbitrary field $\mathfrak{F}$. They imply that a polynomial $f(x)$ of degree t can have at most t distinct roots in the field $\mathfrak{F}$. Thus there are at most p values a in $\mathfrak{F}$ such that $a^p - a = 0$. But there are $p^n > p$ elements in $\mathfrak{F}$, and so $\mathfrak{A}_\delta$ is not associative for every δ, satisfying our condition.

We state the final result as a theorem.

THEOREM 3. The algebraic systems $(\mathfrak{A}_\delta, \oplus, \circ)$ can be used to coordinatize non-Desarguesian planes.

Proof. Define the ternary operation F in $\mathfrak{A}_\delta$ by $F(x, m, k) = (x \circ m) \oplus k$ where the elements are *ordered pairs* of elements of $\mathfrak{F}$, and we must check the conditions of Theorem 4.1. Conditions (a), (b), and (d) are trivially satisfied. For (c), we have $(x \circ a) \oplus b = (x \circ c) \oplus d$. Using the algebraic properties of $\mathfrak{A}_\delta$, this can be rewritten as $x \circ (a \ominus c) = b \ominus d$, which has a unique solution for x, since $(\mathfrak{A}_\delta - \{(0, 0)\} \circ)$ is a loop and $a \ominus c \neq (0, 0)$. Finally, Theorem 4.3 completes the proof, showing that (e) is also satisfied.

As promised, there was constructed in this section an infinite class of finite non-Desarguesian planes (coordinatized by V-W systems). This is only one of the many known classes of non-Desarguesian planes, but perhaps the easiest to construct, given the theorems we had at our disposal. Observe that we already know a great deal about the collineation group of the planes coordinatized by the algebraic systems $\mathfrak{A}_\delta$, since the theorems of Chapter 5 and the algebraic structure of $\mathfrak{A}_\delta$ implies the existence of many central collineations in the plane it coordinatizes.

In the next section we shall explicitly construct geometrically a non-Desarguesian plane. First, some exercises that will help fill in the gaps in the construction of this section follow.

Exercises

3. Verify that $(\mathfrak{A}_\delta, \oplus)$ is a commutative group of order p^{2n}.

4. Use the fact that $(\mathfrak{F} - \{0\}, \cdot)$ is a cyclic group of order $p^n - 1$ to show that if $p \neq 2$, $n > 1$, there must exist an element in $\mathfrak{F}$ which is not a $(p + 1)$st power of any element in $\mathfrak{F}$. (Hint: Show that there is an element $x \neq 1$ satisfying $x^{p+1} = 1$; and then show that the mapping $x \rightarrow x^{p+1}$ cannot be an onto mapping.)

5. Verify the fact that $(\mathfrak{A}_\delta - \{(0, 0)\}, \circ)$ is a loop.

6. Verify both distributive laws in $\mathfrak{A}_\delta$.

7. Show that, for any polynomial $g(x)$ with coefficients in $\mathfrak{F}$, $g(x) = q(x)(x - a) + r$, where the degree of $q(x)$ is less than that of $g(x)$, and r is in $\mathfrak{F}$, for any a in $\mathfrak{F}$.

3. A Concrete Example

A famous result in the subject of finite projective planes says that *any finite plane of order* ≤ 8 *must be Desarguesian.* In this section, we shall construct explicitly a non-Desarguesian plane of order 9. Of course, the planes of Section 2 include one of this order, but the one described here is different, and it is also described more geometrically. In order to define this plane, we first have to know some more details about the field $GF(9)$. Let $\mathfrak{F} = GF(3) = \{0, 1, 2\}$, whose operations are given by:

$$\begin{gathered} a + 0 = 0 + a = 0, \\ 1 + 1 = 2, \qquad 1 + 2 = 0 = 2 + 1, \qquad 2 + 2 = 1 \\ 1 \cdot 2 = 2 \cdot 1 = 2, \qquad 2 \cdot 2 = 1, \qquad 1 \cdot 1 = 1. \end{gathered} \tag{15}$$

Next, let $\mathfrak{K}$ be the set of objects: $\{0, 1, 2, \lambda, 1 + \lambda, 2 + \lambda, 2\lambda, 1 + 2\lambda, 2 + 2\lambda\}$ and define

$$a\lambda = \lambda a, \qquad a + b\lambda = b\lambda + a$$
$$(a + b\lambda) + (c + d\lambda) = a + c + (b + d)\lambda, \tag{16}$$
$$(a + b\lambda) \cdot (c + d\lambda) = ac + bd + \lambda(ad + bc + bd),$$

for all a, b, c, and d in $\mathfrak{F}$.

This defines two binary operations in $\mathfrak{K}$ which makes $\mathfrak{K}$ into a field–-the field $GF(9)$.

Exercise

8. Verify that $(\mathfrak{K}, +, \cdot)$ satisfies the conditions for a field and is thus a representation of $GF(9)$.

To define the plane Π', consider the plane $\Pi = PG(2, 9)$ coordinatized by $\mathfrak{K}$. The plane Π' is to consist of the points of Π, and the definition of Π' will consist of redefining the subsets of Π that are to be the lines (remember that our original definition of projective planes was in terms of sets and subsets).

Several of the lines (subsets) of Π will also be considered as lines (subsets) of Π'. These are $\mathfrak{L}_S$ (the line consisting of Y_S and all points named (x), for x in $\mathfrak{K}$) and all lines $[m, k]$ for m not in $\mathfrak{F}$. A line of this last type consists of the point (m) and all $(x, x \cdot m + k)$. Now, since there are 6 elements in $\mathfrak{K} - \mathfrak{F}$, and 9 lines of the form $[m, k]$ through each, this gives 55 lines of Π', and leaves 36 lines to be defined. The remaining subsets are defined as follows: For any x not zero, and y, z in $\mathfrak{K}$, let $x = a + \lambda b$ (a, b in $\mathfrak{F}$) and define $P_x = Y_S$ if $a = 0$, $P_x = (ba^{-1})$ if $a \neq 0$. We then define the subset $\mathfrak{U}_{x,y,z}$ by

$$\mathfrak{U}_{x,y,z} = \{(x\mathfrak{F} + y, x\mathfrak{F} + z)\} \cup \{P_x\}, \tag{17}$$

where $\{(x\mathfrak{F} + y, x\mathfrak{F} + z)\}$ represents all 9 points that can be obtained by running through all elements of $\mathfrak{F}$ in both places where $\mathfrak{F}$ is written; that is,

$$(xa_1 + y, xa_2 + z) \in \mathfrak{U}_{x,y,z} \tag{18}$$

for any a_1, a_2 in $\mathfrak{F}$.

The proof that this choice of subsets makes Π' into a projective plane requires some groundwork, after which the completion of the proof can be left as a set of exercises.

First, we shall introduce a mapping that will make the proof rather simple. For any $(x, y) = (\alpha + \lambda\beta, \gamma + \lambda\delta)$, where α, β, γ, δ are in $\mathfrak{F}$, we define

$$(x, y)^\sigma = (\alpha + \lambda\gamma, \beta + \lambda\delta). \tag{19}$$

Observe that $((x, y)^\sigma)^\sigma = (x, y)$, so $\sigma \circ \sigma = i$, the identity.

Now, let $\mathfrak{U}_{x,y,z}$ be any subset where $x = 0 + \lambda b$. Then, if $y = y_0 + \lambda y_1$, $z = z_0 + \lambda z_1$, where the y_i and z_i are in $\mathfrak{F}$, we can easily show that

$$\mathfrak{U}_{x,y,z} = \mathfrak{U}_{\lambda,y_0,z_0}. \tag{20}$$

For $(\lambda b f_1 + y_0 + \lambda y_1, \lambda b f_2 + z_0 + \lambda z_1) = (y_0 + \lambda f'_1, z_0 + \lambda f'_2)$, where the f_i and f'_i are in $\mathfrak{F}$. Now, for any point P in $\mathfrak{U}_{\lambda,y_0,z_0}$, $P = (y_0 + \lambda\alpha, z_0 + \lambda\beta)$, for some α, β in $\mathfrak{F}$. Then $P^\sigma = (y_0 + \lambda z_0, \alpha + \lambda\beta)$. But as α, β vary over all elements in $\mathfrak{F}$, the points P^σ vary *over all* elements (not on $\mathfrak{L}_S$) in the line of $\Pi = PG(2, 9)$, which is called $[y_0 + \lambda z_0]$, for all points P^σ have $y_0 + \lambda z_0$ as their first coordinates. Similarly, given any line $[k] = [y_0 + \lambda z_0]$ of Π, it can be made to correspond to the subset $\mathfrak{U}_{\lambda,y_0,z_0}$ of Π'.

Next, we shall consider sets of the form $\mathfrak{U}_{x,y,z}$, where $x = a + \lambda b, a \neq 0$. But $x\mathfrak{F} = (1 + \lambda b a^{-1})\mathfrak{F}$ (Why?), so we can assume $x = 1 + \lambda\alpha$ for some α in $\mathfrak{F}$. Now, for P in $\mathfrak{U}_{x,y,z}$, $P = ((1 + \lambda\alpha)f_1 + y_0 + \lambda y_1, (1 + \lambda\alpha)f_2 + z_0 + \lambda z_1) = (f_1 + y_0 + \lambda(\alpha f_1 + y_1), f_2 + z_0 + \lambda(\alpha f_2 + z_1))$ for some f_1, f_2 in $\mathfrak{F}$. Thus,

$$P^\sigma = (f_1 + y_0 + \lambda(f_2 + z_0), \alpha f_1 + y_1 + \lambda(\alpha f_2 + z_1)) = (x, x\alpha + d), \tag{21}$$

where

$$x = f_1 + y_0 + \lambda(f_2 + z_0) \tag{22}$$

and

$$d = (y_1 - y_0\alpha) + \lambda(z_1 - z_0\alpha). \tag{23}$$

Now, this shows that each subset $\mathfrak{U}_{x,y,z}$ of the type we are examining is mapped by σ onto the points (not on $\mathfrak{L}_S$) of the line $[\alpha, d]$ of Π, for some α in $\mathfrak{F}$. Similarly, given $[\alpha, d]$ a line of Π, for α in $\mathfrak{F}$, we can use σ to determine a subset $\mathfrak{U}_{1+\lambda\alpha,y,z}$ (How?).

These considerations show at least that there are the right number of sets $\mathfrak{U}_{x,y,z}$ to qualify Π' to be a projective plane of order 9. The remainder of the proof consists in checking the axioms for a projective plane, but these follow fairly simply using the transformation σ. For example, if $P = (x, y)$ and $P_1 = (x_1, y_1)$ are two points of Π', then P^σ and $P_1{}^\sigma$ determine a unique line of type $[k]$, or $[m, k]$ of Π. If they determine the line $[x]$ of Π, where $x = y_0 + \lambda z_0$ (where y_0, z_0 are in $\mathfrak{F}$), we see by looking at Q^σ for $Q \, \mathrm{I} \, [x]$, that $[x]$ corresponds to the subset $\mathfrak{U}_{\lambda,y_0,z_0}$ of Π', and P and P_1 uniquely determine this subset of Π'. Similarly, let $[P^\sigma \cdot P_1{}^\sigma] = [m, k]$ in Π. If m is not in $\mathfrak{F}$, then the line $[m, k]$ is also the unique "line" of Π', determined by P and P_1. If m is in $\mathfrak{F}$, then P and P_1 determine one of the subsets $\mathfrak{U}_{x,y,z}$ (Which one?). In this way, we easily show that any two points of Π' determine a unique "line" of Π'.

To show that any two lines of Π' determine a unique point of Π', it suffices (Why?) to show that a line of type $[m, k]$, with m not in $\mathfrak{F}$, and a subset $\mathfrak{U}_{x,y,z}$, have exactly one point in common. To see this, we consider the point (a, b) in $\mathfrak{U}_{x,y,z}$, such that $(a, b)\,\mathrm{I}\,[m, k]$. Then

$$(a, b) = (x\alpha + y, x\beta + z) \qquad \alpha, \beta \in \mathfrak{F}. \tag{24}$$

Now, $(a, b)\,\mathrm{I}\,[m, k]$ if and only if

$$b = am + k, \tag{25}$$

or $x\beta + z = (x\alpha + y), m + k = x\alpha m + ym + k$.

This last equation can be rewritten $x(\beta - \alpha m) = ym + k - z$, or $\beta = \alpha m + (ym + k - z)x^{-1}$ (since $x \neq 0$). Now, $(ym + k - z)x^{-1} = f_1 + \lambda f_2$ for some f_1, f_2 in $\mathfrak{F}$, and $m = g_1 + \lambda g_2$ for some g_1, g_2 of $\mathfrak{F}$, where $g_2 \neq 0$ since m is not in $\mathfrak{F}$. Thus, for fixed f_i, g_i, we have

$$\beta = \alpha g_1 + \lambda\alpha g_2 + f_1 + \lambda f_2 = \alpha g_1 + f_1 + \lambda(\alpha g_2 + f_2). \tag{26}$$

In order for (α, β) to give a solution to (26), and hence to (25), using the fact that α and β are in $\mathfrak{F}$, we must have $\alpha g_2 + f_2 = 0$. Since $g_2 \neq 0$, $\alpha = -f_2 g_2$ and since β is in $\mathfrak{F}$, (26) gives a unique solution for β as well as α. Thus there is a solution, and it must also be unique. This, along with a consideration of the simpler cases, completes the proof that Π' defines a projective plane.

Exercise

9. Complete the cases necessary to prove Π' a projective plane.

Thus far, we have merely redefined the subsets of Π to create the projective plane Π'. Some additional work is required to prove that Π' is not itself Desarguesian (that is, isomorphic with Π). This could be done by coordinatizing Π' and showing that its coordinatizing ternary ring is not a field, but it is simpler to construct two triangles perspective from a point P and not perspective from any line $\mathfrak{L}$. This will be left as an exercise with the following hints: In Π', consider the triangles $(\lambda + 1)$, $(\lambda, \lambda + 1)$, $(2\lambda + 1, 0)$ and (λ), $(\lambda, 1)$, $(1, 2\lambda)$. Then show that these two triangles are perspective from Y_S, by showing that $[(\lambda + 1) \cdot (\lambda)] = \mathfrak{L}_S$, $[(\lambda, \lambda + 1) \cdot (\lambda, 1)] = \mathfrak{U}_{\lambda,0,1}$, and $[(2\lambda + 1, 0) \cdot (1, 2\lambda)] = \mathfrak{U}_{\lambda,1,0}$.

The next step is to show that $[(\lambda + 1) \cdot (\lambda, \lambda + 1)] \cap [(\lambda) \cdot (\lambda, 1)] = (0, 2\lambda)$ and $[(\lambda + 1) \cdot (2\lambda + 1, 0)] \cap [(\lambda) \cdot (1, 2\lambda)] = (0, \lambda)$. After this, we can show that $[(0, 2\lambda) \cdot (0, \lambda)] = \mathfrak{U}_{\lambda,0,0}$. Finally, it must be demonstrated that $[(\lambda, \lambda + 1) \cdot (2\lambda + 1, 0)] \cap \mathfrak{U}_{\lambda,0,0} = (\lambda, 0)$, while $(\lambda, 0) \not\mathrm{I}\, [(\lambda, 1) \cdot (1, 2\lambda)] = [2\lambda + 1, 2]$. This proves that Π' is not $(Y_S, \mathfrak{U}_{\lambda,0,0})$-Desarguesian, so Π' cannot be Desarguesian.

Exercise

10. Perform the indicated calculations which show that Π' is not Desarguesian.

Finally, we discuss an algebraic construction of a linear ternary ring $(\mathfrak{R}, F)$ of order 9 such that $(\mathfrak{R}, \oplus, \cdot)$ is not a field. In fact, it is true that $(\mathfrak{R}, F)$ is one of the ternary rings of the plane defined earlier in this section, but we shall not attempt to prove that here.

To define $(\mathfrak{R}, F)$, let $\mathfrak{K} = GF(3)$. The set $\mathfrak{R}$ will consist of all distinct ordered pairs (a, b), where a and b are in $\mathfrak{K}$. Thus $\mathfrak{R}$ has 9 elements. To define the ternary operation $\mathfrak{F}$, we shall define two binary operations $+$ and $\cdot$, and assume that $(\mathfrak{R}, F)$ is linear. First, define

$$(a, b) \oplus (c, d) = (a + c, b + d), \tag{26}$$

for any a, b, c, d in $\mathfrak{K}$. Next, to define $\cdot$, write

$$(a, b) \cdot (c, d) = (ac - d^{-1}bc^2 + d^{-1}bc + d^{-1}b, ad - bc + b) \tag{27}$$

if $d \neq 0$, and

$$(a, b) \cdot (c, 0) = (ac, bc). \tag{28}$$

Clearly, $(0, 0)$ satisfies $(0, 0) \oplus x = x \oplus (0, 0) = x$ for any x in $\mathfrak{R}$, and $(1, 0)$ satisfies $(1, 0) \cdot x = x \cdot (1, 0) = x$ for any x in $\mathfrak{R}$. Finally, for any x, m, k in $\mathfrak{R}$, we define

$$F(x, m, k) = (x \cdot m) \oplus k. \tag{29}$$

We leave the remainder as a difficult exercise.

Exercise

11. Prove that $(\mathfrak{R}, F)$ is a *planar ternary ring* satisfying (1) $(\mathfrak{R}, \oplus)$ is a group, (2) $x \oplus y = y \oplus x$, and (3) $(x \oplus y) \cdot z = (x \cdot z) \oplus (y \cdot z)$. Next, show that *any* element $x = (a, b)$, with $b \neq 0$, satisfies the equation $x^2 - x - (1, 0) = (0, 0)$, and since there are 6 such elements in $\mathfrak{R}$, that implies that $(\mathfrak{R}, \oplus, \cdot)$ is not a field, because *in a field a quadratic equation can have at most two roots.*

Appendix

The Bruck-Ryser Theorem

One of the most important unsolved problems in the subject of finite projective planes is that of determining whether an integer, n, can be the order of a projective plane. Of course, if n is a power, p^m, of a prime p, there is a field plane of order n. Also, since every finite field has order p^m for some prime p and positive integer m, it is clear that every field plane has prime-power order. It is somewhat more difficult to prove that *every finite* V-W *system has prime-power order*, but very little is known about the most general projective planes.

In fact, only one result in this direction has ever been proved, and that is the famous Bruck-Ryser Theorem, published in 1949 by R. H. Bruck and H. J. Ryser. It will be stated without proof in this exposition. A proof can be found in M. Hall's book,* the result of which is as follows.

BRUCK-RYSER THEOREM. Let n be an integer such that either $n - 1$ or $n - 2$ is divisible by 4. Then if n cannot be expressed as the sum of two integral squares ($n \neq a^2 + b^2$ for a and b nonnegative integers), there exists no finite projective plane of order n.

As a consequence of the Bruck-Ryser Theorem, we see immediately that there is no plane of order 6, since $6 - 2 = 4$, which is divisible by 4, and $6 \neq a^2 + b^2$ for any integers a, b. The next integer greater than 6 that is not a prime power is 10. Now $10 - 2 = 8$, but $10 = 3^2 + 1^2$, so the Bruck-Ryser Theorem gives us *no* information regarding the existence of a plane of order 10. In fact, much work has been done on large computers to attempt to construct a plane of order 10, but none has yet been found and not all possibilities have been exhausted, so the case of 10 remains open. As we look at larger and larger integers, we find many that, like 6 and 14, cannot be the order of a finite projective plane. Indeed, it is not too difficult to show that there are an infinite number of possible orders excluded by the Bruck-Ryser Theorem. Nevertheless, there are also an infinite number (10, 12, 15, $\cdots$) of integers that are neither prime-powers nor excluded by the Bruck-Ryser Theorem, and no projective planes of any of those orders are known; nor has anyone been able to prove that any of these integers cannot be the order of a plane.

* The Theory of Groups, pp. 394–398.

References

We list here a few of the most important expository books and papers on the subject of projective planes.

The American Mathematical Monthly, No. 4 of the Slaught Memorial Papers, *Contributions to Geometry*, **62** (1955).

Bruck, R. H., and Ryser, H. J., The non-existence of certain finite projective planes. *Canadian Journal of Mathematics* **1** (1949), 88–93.

Hall, Marshall, Jr., Projective Planes, *Transactions of the American Mathematical Society* **54** (1943), 229–277.

Hall, Marshall, Jr., *The Theory of Groups*, Macmillan, New York (1959), 346–420.

Pickert, G., *Projective Ebenen* (German), Springer, Berlin (1955).

Ryser, H. J., Combinatorial Mathematics, *Mathematical Association of America, Carus Monographs No*. 14, New York, 1963.

Index